MAGIC OF MINERALS AND ROCKS

DIRK J. WIERSMA

MAGIC OF MINERALS AND ROCKS

Springer-Verlag Berlin Heidelberg GmbH

Dirk J. Wiersma

Title of original version: *Exposure of Mineral and Rock*
published by Inmerc bv, Wormer, The Netherlands
© 2002 Dirk Wiersma/Inmerc bv

ISBN 978-3-642-62251-9   ISBN 978-3-642-18695-0 (eBook)
DOI 10.1007/978-3-642-18695-0

Library of Congress Control Number 2004103199

Bibliographic information published by Die Deutsche Bibliothek
Die Deutsche Bibliothek lists this publication in the Deutsche Nationalbibliografie;
detailed bibliographic data is available in the Internet at <http://dnb.ddb.de>.

springeronline.com

This edition © Springer-Verlag Berlin Heidelberg 2004
Originally published by Springer-Verlag Berlin Heidelberg New York in 2004
Softcover reprint of the hardcover 1st edition 2004

Cover design: E. Kirchner, Heidelberg

Printed on acid-free paper   32/2132/ 5 4 3 2 1 0

The world shown by Dirk Wiersma in his photographs is a world of rocks, rocks varying in scale from spectacular mountains and fantastic rock formations to the mysterious micro-world of stones and minerals. Wiersma observes this world with the perspective of the professional geologist, but with a high sensitivity for visual sensations. In his pictures he expresses his emotional perceptions and by doing so he presents us with images that have an air of fairy-tales and enigmas. He reveals the intrinsic beauty of the realm of rocks, that in everyday life, moulds our physical surroundings.

Rocks are everywhere around us, but the 'lifeless world' of the realm of rocks is the antipode of emotions, of thought and life. Stones are not alive, neither are they threatened by death. From times far past stones are the symbols of eternity: they are affected by nothing, except by geologic processes in the course of millions of years. Rocks will still be there when, a few billion years ahead, the phenomenon Life will have disappeared on Earth, while our planet keeps turning around an extinguished sun as a dark and cold globe of stone. But although rocks are far remote from emotions, sensations and other whims of the human mind, they still have always aroused feelings and even passions in man. Already in the distant past the natural beauty of precious stones, be it crude or worked on, tempted people's vanity and were used as symbols of wealth and power. The geometry and mathematical symmetry of crystals, as well as the shape and architecture of rocks and mountains, are a source of inspiration for artistic achievements. From the earliest beginnings man has attributed magic and mysterious powers to minerals and rocks. Crude rocks formed by nature were worshipped as dwelling places of gods or ghosts. Stone sculptures shaped by nature were regarded as sacred places, where holy rituals were performed. Large unprocessed stones were used to set up menhirs, dolmens and other monuments for the death or as cults of the sun. Medieval alchemists searching for the secret of matter, were convinced that it lay concealed in the mythical Lapis Philosophicus (Philosophers' Stone). And the forever holiest shrine of the Islam faith, to which every pious Muslim hopes to make his pilgrimage, is the large black stone of Mecca, the Ka'aba (most probably a meteorite). And even in our mod-

ern and ever so rational society there is still widespread belief that stone amulets and talismans will guarantee the bearer health and a long life, bring prosperity, avert or attract misfortune, protect against evil, heal diseases or bring them about, or work as an aphrodisiac.

But rocks, besides arousing feelings of spirituality, magic or aesthetics, also play a crucial role in all material aspects of life. The erosion products of minerals and rocks allow for plant growth and thus for life on Earth, including that of mankind. In the geological sciences rocks form the archive in which the 4.5 billion years old history of Earth is stored – much like data stored in the silicon archives of our chip era. The geologist reads this rock archive by means of various scientific methods. Our distant ancestors used caves, formed in rocks by geologic processes, as places to live in or to perform cults. Using colorants derived from finely ground minerals of different color, they made intriguing wall-paintings in the caves, probably with magic objectives. To prehistoric man stones were the only durable material there was to make tools of and weaponry. The first hominids that worked stones into primitive tools appeared already 2.4 million years ago. In the course of time stone working developed into an increasingly advanced technology, providing better and more refined tools. Until man at the end of the Stone Age, about 6000 years ago, started making weapons and tools out of metals – metals that are, incidentally, also derived from mineral raw materials. But there is no doubt that the hominids who first made stone weapons and tools, according to a well conceived plan, stood at the very roots of the whole technological and cultural development of mankind.

In addition to his artistic photographs of landscapes, rock formations and details of rocks and minerals, Dirk Wiersma also strikes us with fascinating images of thin rock sections, using various optical and microscopical techniques. In these pictures, that are reminiscent of abstract compositions, we recognize the professional geologist. However, as opposed to abstract art that renounces a realistic rendering of the world around us and seeks for a new order in form and color, we here look at creations of nature herself, rather than fancies of the human mind. Natura Artis Magistra or, as

said by Shakespeare in The Winter's Tale: Art itself is Nature —
with his photographs Dirk Wiersma confirms in this book
again the veracity of this adage, also for those not initiated in
the science of geology.

Harry N.A. Priem
Emeritus Professor at the University of Utrecht and
Curator at the Artis Geological Museum in Amsterdam

nerals and rocks are the foundation of everything, the first
lid substance to have materialized from the gases and liquids
space. And life, in a way we still do not entirely understand,
rang from minerals and rocks. Right up to the 17th century,
cks and minerals were still being referred to scientifically as a
rm of 'livestock', sprouted from seeds or from a kind of inter-
urse between primary substances.

eologically speaking minerals are natural, homogeneous solids
ade up of two or more elements. They are usually inorganic
d crystalline and over 3500 types have been defined. They are
e basic building material of the earth's crust, the moon and
her celestial bodies. The word mineral is derived from minare,
e Latin term for mining. The word implies 'usefulness', and
deed minerals are the raw materials from which we derive
any things that are essential for our survival.

cks are composed of one or more minerals. Rocks exist in
untless types and varieties, coarsely textured or fine, hard or
ft, dark, light or polychrome. Hot lava, volcanic ash, mud or
nd, but also plants, bones and tissue, all come from rock and
ll become rock again.
though we think of rocks as the basic stuff of which the earth
made, they actually only form the outer crust of our planet.
cks float – literally – on what geologists call the earth's man-
, a hot and viscous substratum, underneath that solid crust.

art from the geologist's outlook, there are many other ways
view minerals and rocks. From an archaeologist's viewpoint,
cks and minerals are at the roots of civilization, the oldest
kens of human culture, including tools, amulets, tombstones
d dolmens, monuments and ornaments. Others treasure min-
ls and rocks for their preciousness or for their healing
wer; as objects of singular beauty or strength, sublimated
m complex geologic processes in the deeps beneath us.

hom do rocks and mountains not awe, who is not fascinated
crystals and gems or intrigued by fossils that speak of life in a
tant past? Strolling along a beach or a riverbed, looking down,
pick up pebbles and marvel at their smoothness and shape.
hich child has not taken home flints and pebbles for display
a shelf in his bedroom, or for a collection in a shoe box or
bottle?

Adri Verhoeven, a Dutch sculptor, thinks of rock as a deeply
spiritual and timeless material: 'When you look at rocks you
only see the eroded outside. When you split a rock you see a
new outside; by polishing it you create yet another outside. You
still never uncover the real inside'.
The artist's experience is not unlike that of the scientist. Every
discovery, each deepening of knowledge leads to new mysteries.

My fascination with rocks and minerals as presented in this
work comes from two converging paths in my life: geology and
photography. One could also say: science and imagination or
impression and expression.

The images presented in this book range over a broad scale,
from the field perspective to the microscopic. 'Zooming in' like
this means venturing into new spaces time and again, full of sur-
prising new forms and abstractions. Fantasy and imagination
gradually overtake all notions of scale and realism.

In the field, rocks are perceived in their natural setting; in mountains and outcrops, in the riverbed, along a rough coast, in quarries. This may be massive granite and basalt or orderly sediments, parallel bedded or graciously folded. Well rounded pebbles and smoothly polished rocks in one place: erosion, rubble
and chaos elsewhere.
Closer up, on the macro-scale, the images become more elu-
sive, the viewers imagination is challenged. Grotesque mock-
landscapes and other miraculous scenes arise in the form of
crystals or capricious mineral growth. Illusions and mirages
materialize on the polished surface of an agate or inside the tiny
cavities and cracks in a rock.
Closer up again, on the microscopic scale, yet another universe
opens up. There is increasing abstraction, finer detail and delicate texture, but also more plan and order. Most of the pictures
on the microscopic scale are made of thin sections, wafer-thin
slices of rock observed in transmitted light. Polarization of the
light brings out the wondrous, spectral colors that result from
birefringence in crystal sections. If we allow our mind to drift
away, we see dendritic crystals of manganese-oxide forming
shrubs and forests; we see how radial crystals of prehnite mimic
bird feathers, how pinpoint stars of tourmaline adorn stamp-
size firmaments, and how garnets are helicoidally deformed into
milky ways!

These images are an affirmation of the authenticity and time-lessness of abstraction, long before man reinvented it in modern art.

Often we discern, on all these different scales, a repetitive order of forms and structures. The anticlines, synclines, faults and overthrusts that are seen in mountain ranges are repeated in microfolds and tiny fractures in rock samples the size of a fist, and smaller. The five and six-sided pillars of basalt and trachyte, as seen in quarries or in the field, bear a deceptive similarity to the tiny, columnar crystals of tourmaline and other minerals. These repetitive structures and dynamic forms reflect the fascinating concept of fractalism, the theorem conceived and worked out by the mathematician Benoit Mandelbrot, and ever so attractively presented in his book The Fractal Geometry of Nature (1983). He demonstrates that not only rhythms but also many forms and other phenomena in nature or technology can be broken up in fragments (fractals) of similar form or can repeat themselves on many scales and in totally different materials or processes.

Numerous are the descriptions, notably in older but also in recent writing, of the mysterious and evocative images rendered in stone, in eroded rock as well as on freshly broken or polished surfaces. And such images were often not regarded as purely accidental, or as just a whim of nature. There had to be more to it, something spiritual and yet beyond human understanding.

An outstandingly passionate view of stones in this respect was that expounded by the French anthropologist and writer Roger Caillois (1913-1978). He writes about the inimitable forms and images in stone in several essays that form a culmination of mystical lyricism in his diverse literary work. In the introduction to l'Écriture des Pierres (1970) he writes of stones: 'that are sought after and revered not for their rarity or expensiveness, but rather for some peculiarity in form, a pattern or color that is as bizarre as it is consequential. Nearly always the fascination is aroused by an unexpected resemblance, which is improbable and yet so natural. Stones anyway have something grave, something unyielding and extreme, like being imperishable or already having perished long ago'.

It is evident that stones can be looked at from different per-spectives, and the same is true for the pictures in this book. There is the impressionist viewpoint, that of science and theme. And the other perspective is that of expressionism, of form and beauty. I am a geologist by education and without that background it would not have been possible to render this theme in all its depth and richness. But it is the expressionist perspective, that of the 'art' in minerals and rocks, that primarily inspired me to make these exposures.

Art? Whose art...? Our habitual concept of art is one of objects created by our fellow men and rendered with a pen, a brush, a chisel or a camera. But surely, some- where far outside our field of view, an anonymous artist of peerless supremacy has been at work, a master of unlimited means and boundless inspiration, who created things of beauty in the space and matter around us long before we existed, long before we came with our human concept of art.

Dirk J. Wiersma

The plates in this book are distributed over nine chapters. This arrangement has not been made according to a scientific or other strict order. Rather, my intent has been to start at some distance, with the more familiar, external aspects of mineral and rock and proceed from there to the deeper interiors of the subject.

Each chapter begins with a short introduction, followed by brief explanatory texts for all the plates in that chapter. Page number and title identify each plate. Please note that the explanatory text is a few pages back from the actual plate.

The size of the smaller objects is usually specified in the text, either by referring to something familiar, like a golf ball, or else by giving the actual size in millimeters.

PICTURE FACING TITLE PAGE: COLD FIRE

As outlined on page 60, the interior filling of geodes can be, partly or fully, a precipitate of chalcedony from aqueous solutions that flow into cavities in volcanic rock. The solution often percolates into
the cavity through one little duct. That event is here illustrated in graceful curves. Actual size 40 x 50 mm.

PICTURE ON FRONT COVER: PYRITE (PAGE 53), WOLFRAMITE (PAGE 56), CALCITE ON FLUORITE (PAGE 59), GEODIÑO (PAGE 76), BOTSWANA AGATE (PAGE 90), MALACHITE (PAGE 108/109)

This first chapter shows rocks on the large scale, how they adorn the earth's surface in mountains, cliffs, plains and valleys. They look as if they are timeless and everlasting, but they are, in fact, transient, as ephemeral as everything in life. Mountains and valleys come about by action and movement, tectonic uplift and persistent erosion. Today's mountain chains arose from yesterday's oceans and plains. Over tens of millions of years these mountains will wear down, to become plains or to be swept over by oceans once again. Earth has always been and will always remain a dynamic system, moving to and fro between creation and destruction, order and chaos.

ZABRISKIE POINT The geomorphology seen here, from a well-known viewpoint in Death Valley, California, is called 'badland erosion' and is a typical erosive pattern for deposits of shale, rhyolite or other soft sediments in arid regions, where rain only comes in occasional downpours.

LANDMANNALAUGAR Iceland is where one becomes aware of the making of new land, of new mountains, rocks and minerals; as close to creation as can be. Landmannalaugar, in the southern interior of Iceland, is an area of active volcanism. The striking landscape of elongated, sharply crested hills and V-shaped valleys, is formed by rapid erosion of the soft, rhyolitic lava's that were recently deposited here. Rhyolite is a characteristic type of extrusive rock, formed from shallow magma (molten crust) and rich in silica, aluminum, potash and feldspar. It has subdued, earthy colors, fluctuating from green to yellow-beige and brick red. A recent outpouring of obsidian, a glassy and fluid variety of lava, can be seen meandering across the rhyolite hills like a black snake. A brief classification of rock types is presented in the next chapter.

BRYCE CANYON Nowhere the phenomenon called erosion is more spectacularly typified than by the bizarre, pinkish brown pinnacles and spires of Bryce Canyon National Park, USA.

DER HOHE DACHSTEIN Frost and thaw impart striking features to the bedded limestone that makes up the southern face of Der Hohe Dachstein, the godfather of the Dachstein Massif in the northern Alps of Austria.

RECENT MOVEMENTS This would-be lunar landscape is the North Cainesville Mesa, east of Capitol Reef Park in Utah,

U.S.A. The flat-topped mountain in the background, Factory Butte, is an exemplary butte, also referred to as a mesa. The ridges showing an escarpment on one side and a gentler slope on the other side are called cuestas. The crude and edgy geomorphology exhibited here is indicative of the recent, tectonic movements that are still going on in this area.

CLIFFS OF MOHER Pounded upon relentlessly by mammoth Atlantic waves, the Cliffs of Moher, in western Ireland, remain unperturbed in their dark and menacing steepness, thinly veiled by a diaphanous haze of seawater spray.

KRA KRA Awe-inspiring is the natural scenery of the Vesterålen, an island group off the coast of northern Norway. The mountains here may stand out in crystal-clear light one moment, to become dark and looming within the briefest spell of time, shrouded in clouds.

FJARDARA CANYON Black tuff or tuffstone, formed by the accumulation of ashes and coarser particles from volcanic outbursts, characterizes much of the landscape in South Eastern Iceland. The narrow, winding Fjardara Canyon has been cut in these sediments by a small river. In summertime the crystal clear stream, fringed by colorful vegetation, vies with the steep black canyon walls and strange pinnacles to make this place the setting of an uncanny fairy tale.

WHITE SANDS DESERT This unique desert of glistening gypsum sand, surrounded by dark mountain ranges, is located in southern New Mexico, USA. Wandering into this blinding expanse of rolling dunes is a mesmerizing experience. There is no more than a lone yucca stalk or other lean shrub to remind one of being on planet Earth.

SHIPROCK Shiprock is a striking landmark in northwestern New Mexico, USA, protruding 500 meters up out of the surrounding flat plains. Shiprock is a volcanic neck, the central basaltic cone that remained after the softer slopes and sediments surrounding the ancient volcano disappeared by erosion. Several dykes of the same basaltic rock, one of which is seen in the picture, fan out from the central cone into the surrounding plains. That form has inspired the native inhabitants of this region, the Navajo Indians, to give this rock the much more telling name Tse Bidahi, meaning 'Winged Rock'.

THE LAST SNOW The last patches of snow cling for dear life to this mountain slope of black tuff, near the Vatnajökull glacier in eastern Iceland.

ABANDONED BASALT QUARRY Nature slowly resumes its role in this abandoned basalt quarry near Kasbach, in the German Eiffel region. Basalt is an extrusive, igneous rock, which means a rock solidified from lava that has ascended, by volcanism, to the earth's surface. Upon cooling the basaltic lava-body shrinks, and then often sets in a pattern of elongated, five or six-sided columns. This geometrical form makes these basalts of perfect use for cladding of dykes and other waterfront constructions and in the Eiffel region columnar basalt has been exploited for this purpose since centuries, nearly to the point of depletion. The next chapter also shows some striking examples of such columnar basalts.

DOLERITE Dolerite is an 'intrusive' rock, meaning an igneous rock that solidifies underneath the earth's surface. It is akin to basalt, which however is an 'extrusive' rock, a rock that flows out over the surface. The dolerites of the Pilbara region in West Australia, as seen in this picture, form protrusions and elongate dykes that jut out from the flat plains over large distances, their deep-red crests tufted by softly green tussocks of spinifex grass.

ZABRISKI POINT

LANDMANNALAUGAR

BRYCE CANYON

DER HOHE DACHSTEIN >

CLIFFS OF MOHER

KRA KRA

FJARDARA CANYON ›

< WHITE SANDS DESERT

SHIPROCK

THE LAST SNOW

ABANDONED BASALT QUARRY

DOLERITE

Here we see rocks in the field, in natural outcrops, in man-made quarries or the stonemason's yard. Rocks are infinitely diverse, becoming even more so the closer one looks. Geologists have brought as much order in this diversity as possible, starting with a subdivision in three major categories: igneous, sedimentary and metamorphic rocks.

'Igneous rocks' could be considered as the primary rocks on earth, formed by the solidification of magma in the deeper reaches of the earth's crust or of lava that has ascended to the surface. Examples of igneous rock are granite, dolerite, basalt and rhyolite.

'Sedimentary rocks' are those deposited in oceans, lakes, deserts, or other parts of the earth's surface by the activities of water, wind, ice, or chemical precipitation. Sedimentary rocks are usually bedded, layer-by-layer, and capable of reaching massive thickness. A few examples are shale, sandstone, limestone and rock-salt.

'Metamorphic rocks' are igneous or sedimentary rocks that have been altered into a totally different rock by high pressure and temperature, as well as by chemical processes. This happens after such rocks have been buried at appreciable depth below the earth's surface. Limestone, for example, is altered into marble, sandstone into quartzite, claystone and shale into slate or schist, etc.

All these rock types can be further subdivided into a highly differentiated array, each with their own characteristics, each telling their own story of times long past. Geology is ever so creative!

THE GIANTS CAUSEWAY These columnar basalts along the Antrim coast of Northern Ireland are part of a large series of extrusive rocks, straddling the Northern Channel and reappearing along the western Scottish coast, where a renowned outcrop is Fingal's Cave on the Isle of Staffa. Legend has it that the Irish giant, Finn Mac Cool, created the Causeway as a suite of steppingstones for an invasion of Scotland. The Causeway is only one of many examples that exist all over the world, of a striking outcrop or rock being the subject of superstition and ancient myths.

COASTAL BASALT Late afternoon sunlight falls on the majestic basalt columns in a quarry in the German Eiffel region, where such columns are still actively exploited for various applications in construction.

COLUMNAR BASALT Softly curved basalt columns and smoothly rounded boulders create a deceptive air of stillness and kindliness near Framnes, on the North Icelandic coast.

TOTEM Like an indestructible totem pole, hewn from granite, this rock stands watch on the stormy ocean coast of Torndirrup near Albany in southern Australia.

PARTITION The drama of erosion is well exhibited in this scene in the Petrified Forest Park, Arizona, USA. Long ago, a huge chunk of sandstone parted from its massive, maternal formation to start a life of it's own. After being worn and rounded by wind and water, that very chunk falls apart by the same forces and weaknesses by which it originally came loose.

BEEFY BEACH This assembly of pebbles, interspersed on a bedrock of beefy-red jasper, along the seashore of Point Samson, West Australia, is derived from some of the oldest dateable formations on earth. Their age is in the range of 3 billion years!

GRANITE GRACIOUSLY Granite and gneiss make up large stretches of the southwestern coast of Australia, like here at Cape Leeuwin on the very southwestern tip of the continent. These rocks are gorgeously sculpted and molded by the pounding waves of the Indian Ocean. Standing here, staring out, one may see a ship or the occasional whale, but beyond that there is nothing, nothing but ocean for thousands of miles, until the chaste and chilly whiteness of Antarctica dooms up.

THE RED SOURCE Iron gives an arresting red tint to a travertine deposit precipitated from a small spring in Kirkjufell, a volcanic region in southern Iceland. The snow has just barely melted, as can be told from the bleached green of the grass fringing the spring.

GRANITE GRACIOUSLY Granite is usually thought of as a rather forbidding rock, massive and grey. But that is not always true. Here, in Ploumanach, in northern Brittany, France, the sea and other elements have played about with this crude and austere rock, sculpting it into gracious, if not lascivious forms.

JAWS These menacing jaws only seemingly disrupt the serene harmony of the famed, bone-white travertine deposits of Pamukkale, Turkey.

DESERT TOMBSTONES Like tombstones in a deserted graveyard, these karstified limestone remains emerge from the barren, yellow dunes of Nambung, West Australia. See also next plate.

THE PINNACLES These are not megaliths, erected by an ancient tribe to honor their gods, but natural limestone pinnacles in the yellow sand dunes of Nambung, West Australia. They are the remains of limestone beds that were karstified, which means they were leached out by water and further eroded. But surely these pinnacles were, to the aboriginal people of Australia, as sacred and significant as megaliths. Every striking rock or rock formation in their homeland is considered a holy and essential object, marking the songlines and dreampaths devised by their ancestors. Unlike ancient tribes in other parts of the world, the Australian aboriginals felt no need to upend or move rocks around in order to corroborate their beliefs.

CARRARA MARBLE Wandering among the marble quarries, in the Apuanian hills above Carrara, Italy, feels like walking in a labyrinth of gigantic fridges. On Sundays, when no work is done, an uncanny silence reigns, until a dog barks or a stone falls. These sounds reverberate from all sides, for a long time. The softly white marbles quarried here have provided for the needs of builders and sculptors alike since classical times.

THE EMBRACE Those who cannot afford the works of sculpture that they fancy, or who are unimpressed by what artists make these days, can go to Harleville or several other quarries in the area south of Paris. At the interface of certain sedimentary beds one can find strange looking, rounded concretions of fine-grained, white sand, tightly cemented by past diagenetic processes. To local quarrymen and geologists these nodules are known as gogottes or poupées, names that are suggestive of the bulbous, if not lascivious forms that these natural works of art can adopt. Actual height 350 mm.

A BURREN MONUMENT The Burren, on the Irish west coast, is a remarkable land of bare limestone, scarcely overgrown, as rocky as can be. Ice Age glaciers first levelled this terrain out into large flat stretches and smoothly rounded hills. The downpours and general wetness that are ageless in this coastal region next made the limestone prey to intense leaching and karstification. These and other erosive forces transformed the surface into endless stretches of highly fragmented and fissured pavement. In spite of its stony crudeness, this realm – since primeval times – has attracted people who have managed to scrape a living from the inhospitable soil. The people of the Burren had a strange affinity with these rocks. They used them not only for building their houses, but also heaped them up in endless dry stonewalls and other constructions of dubious functionality. Of no rock or heap of stones in the Burren one can be sure whether it is in its natural position or moved about by people, somewhere back in time. The same is true for the upended chunk of karstified limestone seen in this plate.

SVARTIFOSS Svartifoss is a legendary waterfall in Skaftafell Park, southeastern Iceland. Svartifoss means 'Black Falls'. Indeed, this damp setting, with its menacing black basalt columns, always shrouded in a thin spray, with water trickling all over, is strangely disquieting, as if it is not meant to be visited by ordinary mortals, let alone noisy tourists.

COLUMNAR BASALT

COASTAL BASALT

BEEFY BEACH

BREAKING THE WAVES >

THE RED SOURCE

GRANITE GRACIOUSLY

JAWS

< DESERT TOMBSTONES

THE PINNACLES

<A BURREN MONUMENT

Crystals are the most lucid testimony to the strict order that exists in the domain of rocks and minerals. The form of a crystal, that is the arrangement of its faces and angles, is rigorously based on the form and arrangement of the atoms forming that crystal. With that, crystallization handsomely illustrates the theorem of fractalism, referred to in the prologue.

Many of the minerals and rocks on earth are crystalline. Crystals are omnipresent, although they do not always manifest themselves by regular, geometric forms and lustrous facets. One might say that, for the lithosphere, crystallization is as essential as cell growth is for the biosphere. Perhaps the two are not as far apart as we often intimate, prejudiced as we are in our strict separation of the organic and the inorganic, of life and death.

DOMUS This group of scalenohedric calcite crystals, illuminated from underneath, might be a mock-up of one of these prestigious, all-glass buildings, at night in a deserted city. Actual width 140 mm.

QUARTZ Quartz is the mineral silica (SiO_2) in its crystalline form. Quartz is, after feldspar, the most common rock-forming mineral of the Earth's crust. Quartz is also synonymous with brilliance and glitter, if not with magical power. Quartz is the ultimate embodiment of crystallinity. Nearly all minerals, and many synthetic materials, occur as crystals or form crystals, but none on the scale and with such variety of forms as quartz. Actual size 50 mm.

ANDESITE Twinned crystals of feldspar float in a finely crystalline matrix, like lost spaceships. Andesite is a volcanic rock in which such a divergence between large crystals (phenocrysts) and a finely crystalline matrix is typical. This texture is called porphyritic and it results from 'exsolution', a process of separation that takes place during successive phases of crystallization in the lava. This picture is of a thin section exposure (for explanation see page 112). Actual size 20 x 26 mm.

FAYALITE Fayalite is the iron rich member of the olivine mineral suite, forsterite being the magnesium rich member at the other end of this suite of iron-magnesium silicates. Most natural olivines are a mixture of the two. This specimen however is nearly pure fayalite, in an rare pattern of tabular crystals, covered with symmetrical striations (see page 113) Actual size 50 x 65 mm.

Crystals grow by the layer-upon-layer precipitation of sheets of atoms, from a surrounding smelt or solution on crystal planes already formed. The growth of a crystal may be arrested, for longer or shorter periods, because the supply of new molecules from the surrounding source is interrupted by changed circumstances. Full-grown crystals often exhibit traces of intermediate crystal planes resulting from such interruptions. These are appropriately called ghost structures or phantoms, as seen here in a group of amethyst crystals. This exposure is of a thin section (for explanation see page 112), with an actual size no larger than 15 x 25 mm.

EBULLITION Facetted crystalline spheres of pyrite like these, up to the size of a golf ball, are found in black slates in the area around Millstream, West Australia. Retrieving these fascinating nodules is only for the real 'rock hounds'. The slates are hard to come by, as they occur underwater, in creeks and caves. The slabs retrieved then have to be cleaved in a special way before one knows whether they contain any such nodules or not.

WIDMANNSTETTERSCHE FIGUREN Widmannstetter structures are a distinctive pattern of intertwined crystals in iron-nickel meteorites, made visible by chemical etching of a polished surface of such meteorites. They are named after the first man to study them, Aloys von Widmannstetter, scientist and director of the Technical Museum of Vienna in the early 19th Century. The structures seen in this plate are in a piece from the Gibeon, a well-known, large meteorite that fell in the Namibian desert in prehistoric times. Because this meteorite hit the earth's surface at a relatively flat angle, it rebounded in many small pieces that were strewn around over a large area. Since these chunks did not penetrate deeply into the soil and were well preserved, many pieces were and still are retrieved by man. Ancient African tribes used metal of the Gibeon for weaponry and other utensils. For them this metal was godsend, since it would not rust, which, as we now know, is thanks to the nickel content. Actual size 95 x 120 mm.

PYRITE Pyrite is iron-sulphide, a common and widespread iron-mineral. It belongs in the cubic crystal system, and well-developed crystals of pyrite can be perfect cubes (hexaeders) with smooth metallic surfaces. Romantics and dreamers are known to mistake glossy grains in freshly broken rock for gold and pyrite's nickname is therefore 'fool's gold'. Actual size 70 x 110 mm.

AQUAMARIN The hexagonal crystals of aquamarine have a neat, columnar arrangement in this group. Aquamarine is a variety of beryl and is often found in pegmatites, complex rock suites formed at great depth, often the source of ores and semi precious stones. The micaceous mineral surrounding the crystals is muscovite, also typical for pegmatite. The color of aquamarine varies from green to blue, like the waters of the sea, hence its name. Actual size 50 x 65 mm.

REFLECTIONS FROM OUTER SPACE A set of parallel crystal planes, in a group of rhombic calcite crystals, reflects the light of a sparsely clouded, blue sky. Actual size 50 x 70 mm.

MIMICRY What looks like butterfly wings or the feathers of an exotic bird is actually a thin section exposure of prehnite. Prehnite is an inconspicuous, soft-green mineral, forming radial crystals inside voids in igneous and metamorphic rocks. In thin section with transmitted polarized light, the radial crystals of this mineral take on high order refractive colors. Actual size 18 x 22 mm.

TURMALINE Harpoon-like, this large turmaline crystal punctures a matrix of chlorite, which is deposited in the form of small, rosette-shaped aggregates. Thin section, transmitted polarized light. Actual size 15 x 20 mm.

WOLFRAMITE Wolframite is the principal ore for tungsten, a metal used for the manufacture of lamp filaments and other industrial products. The gleam of spectral colors results from a wafer-thin veneer of chalcopyrite, which covers the tubular crystals of the wolframite. Actual size 20 x 25 mm.

AUGITE This is a thin section exposure, in polarized light, of a cluster of augite crystals. Augite is a mineral from the pyroxene group and the individual crystals typically show parallel bands of internal zoning. Clusters of crystals like these grow out of a central nucleation point and they usually occur in potash rich lava's. The actual size of this crystal is no more than 8 mm.

CALCITE GLOBULES Calcite is the crystalline variety of calcium carbonate, a mineral that occurs in many forms. Shells, corals, limestone and marble are also calcium carbonate. The globules seen here are formed by radial, crystalline growth in a void inside lava and they are a rarity. Actual size 40 x 60 mm.

CALCITE ON FLUORITE Like a stranded space vehicle, this double-ended crystal of calcite has nested itself on an earlier formed group of fluorite crystals. Quartz was just characterized as the 'ultimate embodiment' of crystallinity, but calcite comes a close second, as seen in other examples in this chapter. Actual size 80 x 90 mm.

< ANDESITE   FAYALITE

PHANTOMS

EBULLITION

< WIDMANNSTETTERSCHE FIGUREN
PYRITE

MIMICRY   AUGITE >

TURMALINE

WOLFRAMITE

< CALCITE GLOBULES

CALCITE ON FLUORITE

Geodes are more or less spherical nodules of chalcedony with an interior that may be hollow or (partially) filled with quartz. Quartz and chalcedony are, respectively, the crystalline and amorphous (non-crystalline) variety of silica (SiO_2). The chalcedony in geodes may also be referred to as agate, but that term is meant to indicate a chalcedony of semi-precious quality, discussed in the next chapter. Flintstone, chert and silex are other denominations for chalcedony.

The origin of geodes is explained by the precipitation of silica (or more rarely some other mineral) from circulating subsurface waters, inside hollow spaces in volcanic rocks. These hollow spaces are usually the congealed voids that have remained from gas bubbles that originally occurred in the hot lava. The precipitation of silica from percolating water or vapor is a slow process, responding differentially to various circumstances, like concentration and temperature. By these processes imaginative and colorful scenery emerges, with combinations of banded chalcedony, crystalline quartz, the occasional rosettes of hematite (iron oxide) and dendrites of pyrolusite (manganese oxide). Sawing geodes in halves or in slices, and polishing the cut surfaces, brings out all this imagery.

A place known – already since ancient times – for its geodes, agates and other beautiful minerals is Idar-Oberstein in West Germany. Famous makers of jewelry and objects in stone, for example Fabergé, came here to find their materials. Although most of the indigenous resources have now been depleted, Idar-Oberstein has carried on its fame as the worlds 'epicenter' for precious and semi-precious stones. Numerous workshops where stones are fashioned into jewelry and trinkets as well as many shops and several museums make it a lively business center and a collector's paradise.

LE ROUGE ET LE NOIR Waferthin laminae of intensely red hematite surround a mysterious, black hole of quartz crystals. On the polished face of an Estérel geode the size of a golf ball, cut in half. Geodes of this type, of which we see a few more in the next plates, come from the Estérel region on the French Mediterranean coast and they are quite characteristic. The surrounding greenish-grey host rock is rhyolite, a volcanic rock. It encloses a core of chalcedony, stained by the vivid reds of hematite. To the French geologists who studied these geodes they are also very special. So special that they did not simply call them geodes but rather assigned them a more dignified name, 'Lithophyses'. Also, these scientists worked out a deviant

theory for the origin of these lythophyses, very complex and difficult to comprehend for the outsider. Fortunately, as with many other things of beauty in this country, these precious little gems can be more than adequately enjoyed without complicated explanations.

APOCALYPSE The portentous scenery inside this rhyolitic geode is no bigger than a thumbnail. See above for a more detailed description.

THE AUSTRALIAN ROOSTER This rooster, in a rhyolitic geode (see above) from Australia, stands face to face with his alter ego, from France. Is this coincidental? Or is it a conspiracy, something beyond the understanding of plain mortals? Actual width 35 mm.

LE COQ D'ESTÉREL This proud rooster stands in a small, rhyolitic geode (see above) from the Estérel region in southern France. For some inscrutable reason the precipitation of chalcedony, inside rhyolitic geodes, often takes on the outlines of poultry. Like the crowing rooster on the facing page, flown in all the way from Australia. Actual diameter 40 mm.

ESCAREL This tasteful agate, the size of a plum, is also from the Estérel Mountains on the French Mediterrannée. The surrounding green rock is rhyolitic lava. The void inside has been filled up with clear chalcedony, which is full of little red rosettes of hematite, an iron oxide. If this were a dinner course prepared by Bocuse, it would be called Escargots à l'hématite dans une croûte de rhyolite.

THE BIG BANG A slice of agate, from a Brazilian geode not larger than a peach, is depicted here in transmitted light. The German craftsman who cut and polished this slice saw nothing less in it than the Urknall.

GEODE IN THIN SECTION A thin section, photographed with transmitted, polarized light (see page 112), shows the complex details of precipitation and crystallization inside a geode, with banded rings of chalcedony on the periphery and a core of crystalline quartz. Actual size 15 x 24 mm.

LITTLE ICE CAVE PHANTASY You have to see it to believe it. In western New Mexico, in the inaccessible terrain of black, rugged lava called Malpais (Spanish for 'bad land') there are caves, just under the surface, containing eternal ice, even in midsummer, when surface temperatures rise over 40°C. The scene in this tiny slice of agate, fiery red chalcedony on the outside and small quartz crystals on the inside, is an allegory to this whim of nature. Actual size 45 x 60 mm.

GRADES OF SILICA This polished slice of a geode nicely shows the gradational change from amorphous silica (chalcedony) precipitated in the outer rings, to crystalline silica (quartz) filling the interior. Actual width 55 mm.

BUDDING As outlined in the introduction to this chapter, the interior filling of geodes can be, partly or fully, a precipitate of chalcedony from aqueous solutions that flow into cavities in volcanic rock. The solution often percolates into the cavity through one little duct. That event is here illustrated in graceful curves. Actual size 30 x 48 mm.

FATA MORGANA Two thin slices of agate, cut from different geodes and placed on top of each other, create this illusory water hole. Actual size 40 x 50 mm.

THE DOLPHINS Opposite halves of an Australian geode, the size of a small melon, are suggestively portrayed against a marine background, evoking a scene that can often be seen in the seas around that continent. This geode is formed in a void inside a volcanic rock called rhyolite, the same way as described for the first few pictures in this chapter.

PHONY BLUE The forms taken on by chalcedony inside geodes are wondrous and so are the colors, but…, not always. If the colors fall short, well, there are worse things that can happen. After all, by immersing geode slabs in color baths, they can be given the fanciest colors you wish. Which is what accounts for the wondrous blue in this specimen. Actual diameter 70 mm.

GEODIÑO Geodiño is the term used in South America for small geodes, which often show fine detail, with small and very clear quartz crystals in the hollow core. This one is about the size of a walnut.

A CRYSTAL WALHALLAH The split and polished surface

of a geode, barely the size of a fist, calls to mind the setting of an ancient Germanic saga. Actual size 60 x 90 mm.

SEPTARIAN MAN Septarian nodules are more or less similar to geodes. They are round claystone concretions with an interior structure that is star shaped or formed by haphazardly radiating cracks filled with another mineral, often crystalline calcite. The cracked structure comes about by the drying up and shrinkage of the clay concretion and the cracks and voids are later filled up with calcite. Cut in half and polished, these nodules display intriguing configurations, like the chap valiantly marching ahead in this specimen. Actual diameter 120 mm.

AUSTRALIA This type of geode is called a 'thunderegg'. The colors of the chalcedony filling it are distinctive of the landscape and rocks of the Australian outback. No wonder that the aboriginal people of Australia picked these colors for their national flag. Actual diameter 120 mm.

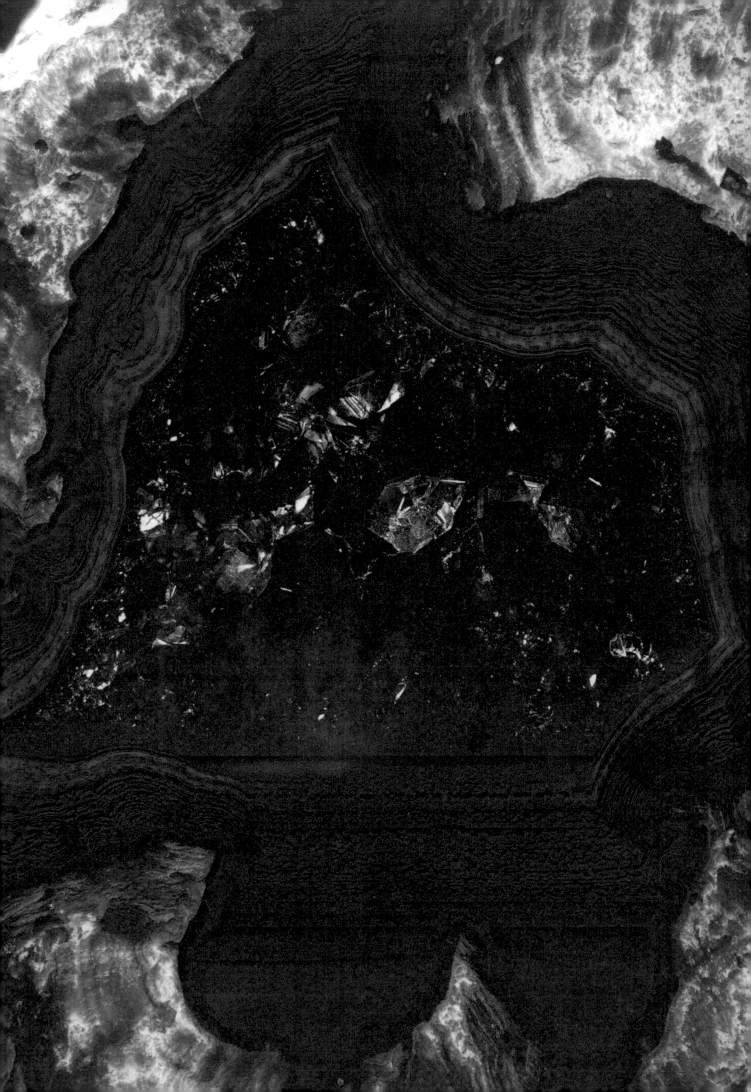

< LE ROUGE ET LE NOIR

APOCALYPSE

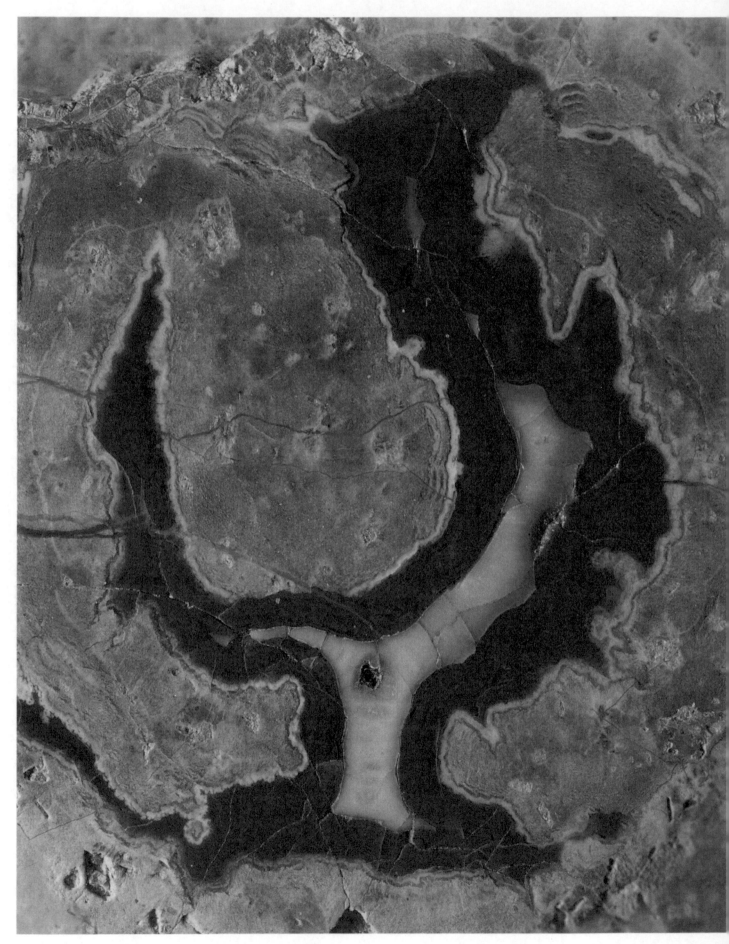

THE AUSTRALIAN ROOSTER

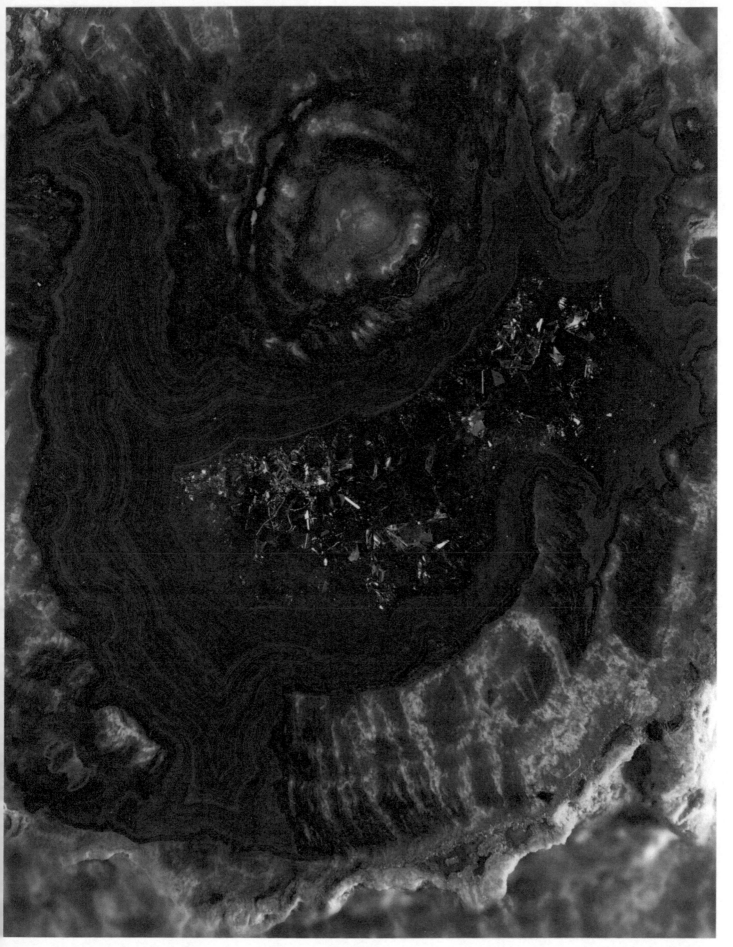

LE COQ D'ESTÉREL

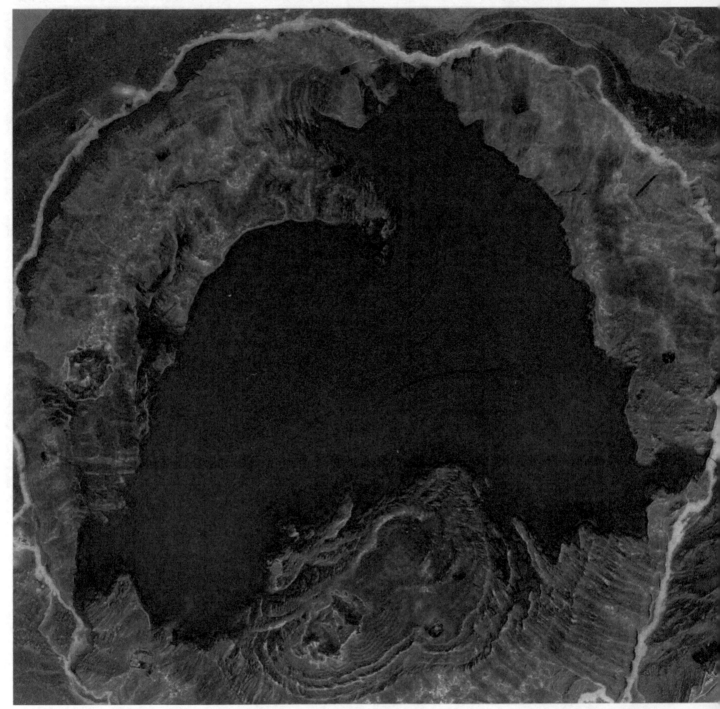

ESCAREL

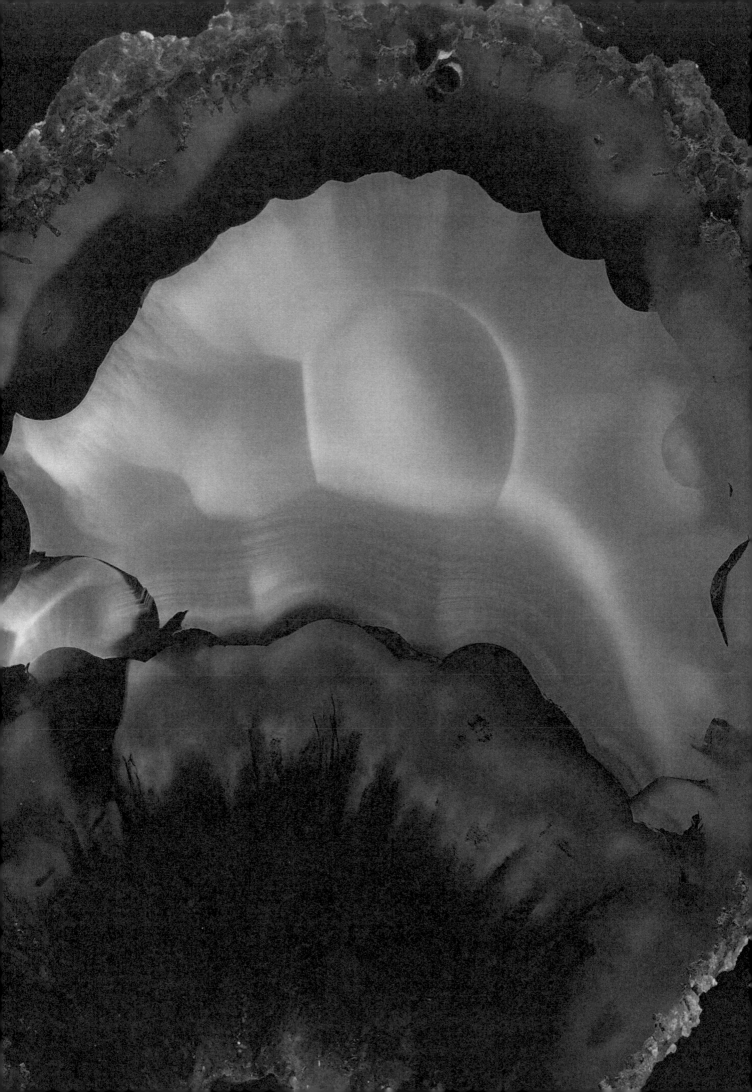

GEODE IN THIN SECTION

LITTLE ICE CAVE PHANTASY

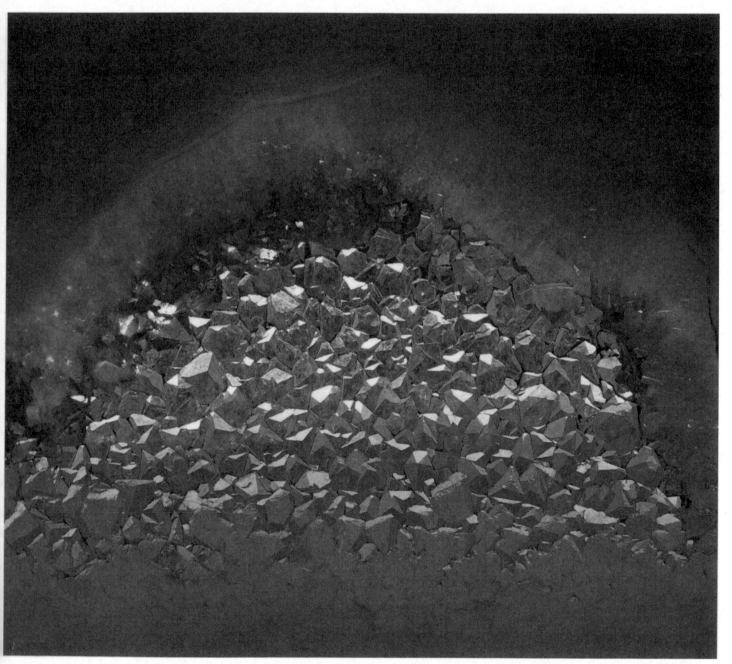

GRADES OF SILICA

BUDDING

FATA MORGANA >

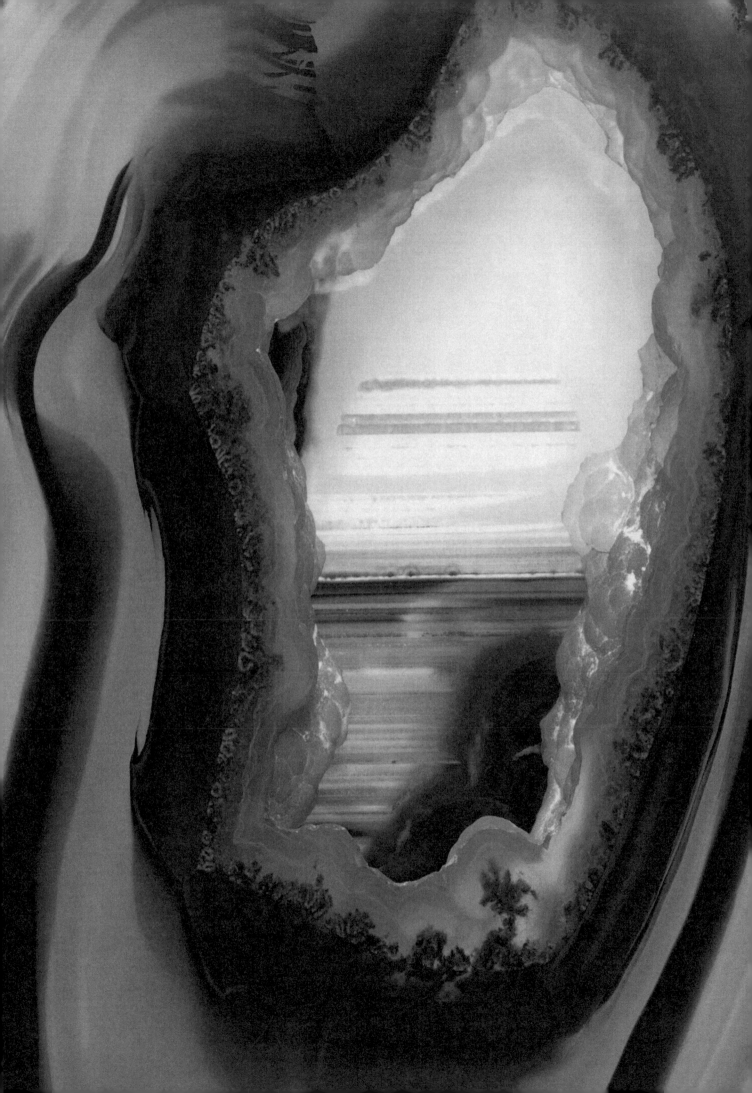

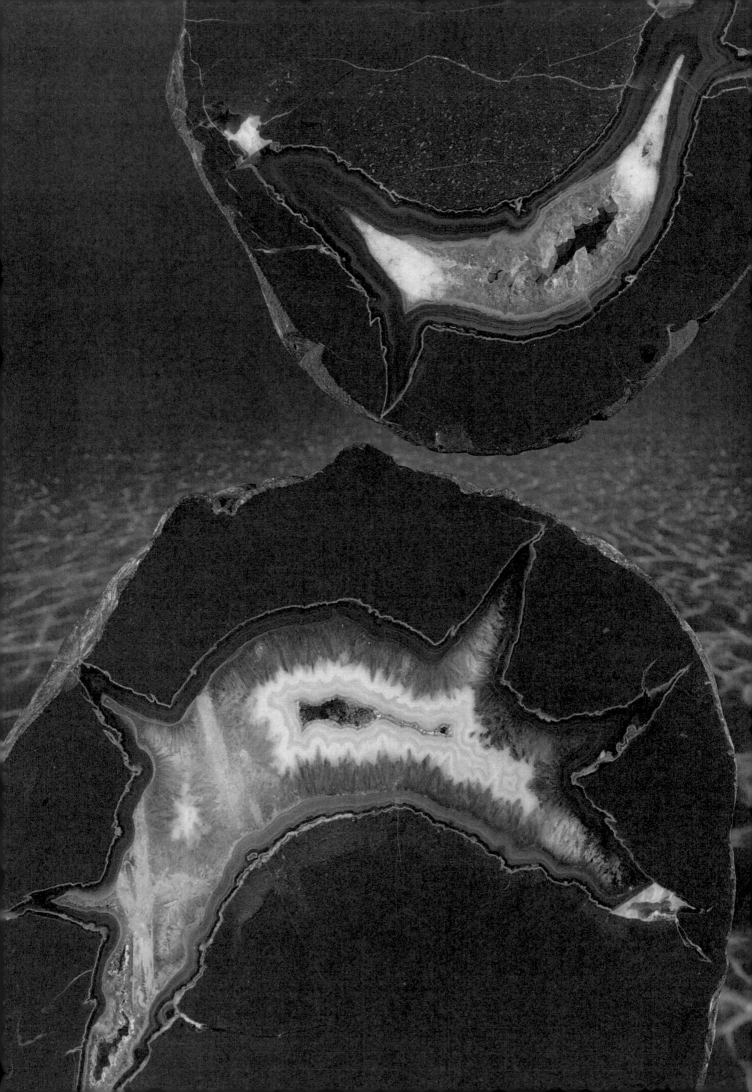

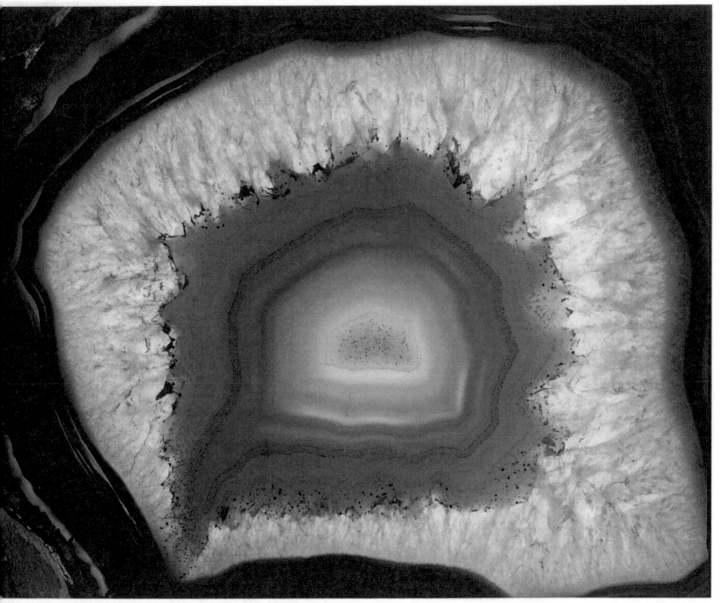

< THE DOLPHINS

PHONY BLUE

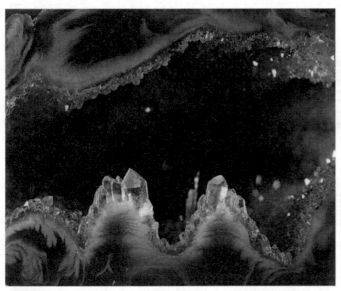

GEODIÑO   A CRYSTAL WALHALLAH >

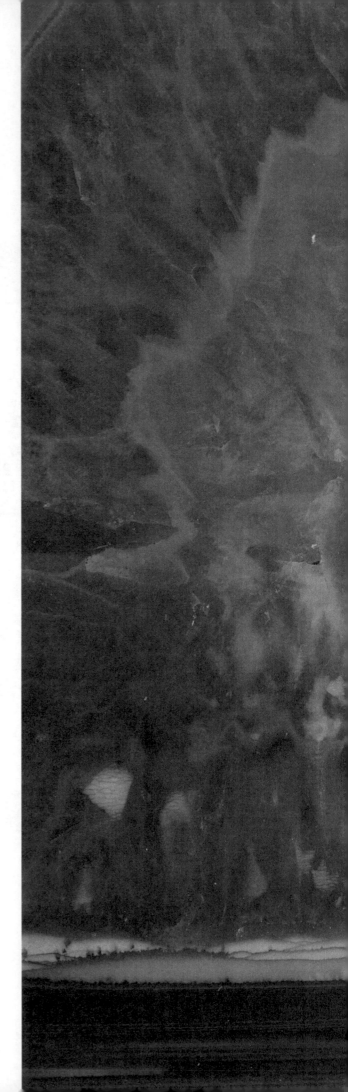

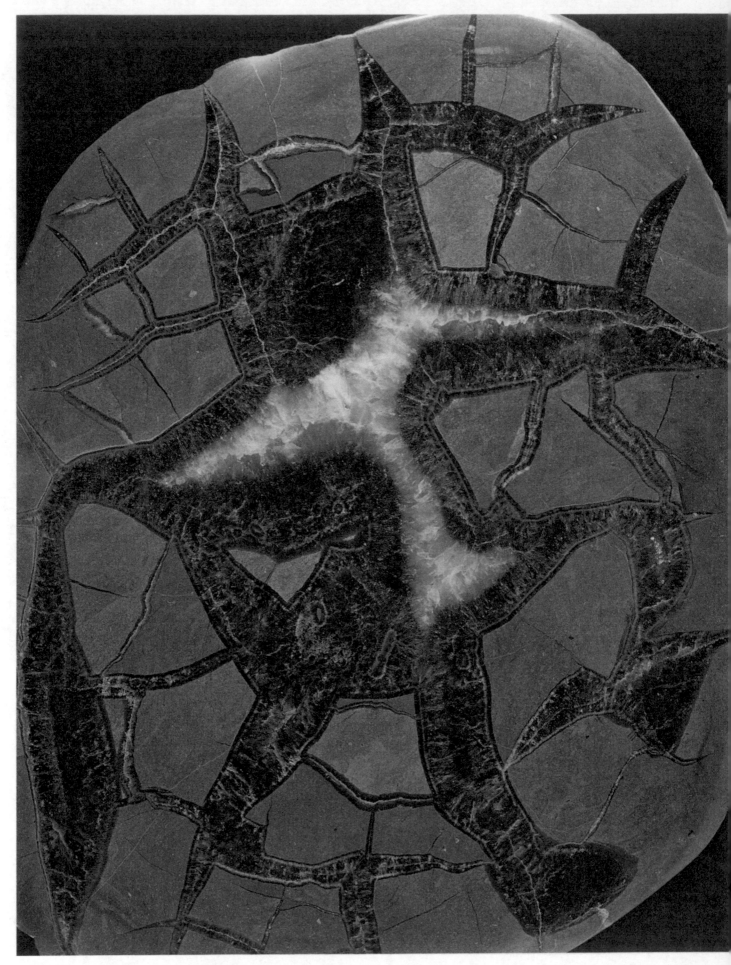

SEPTARIAN MAN

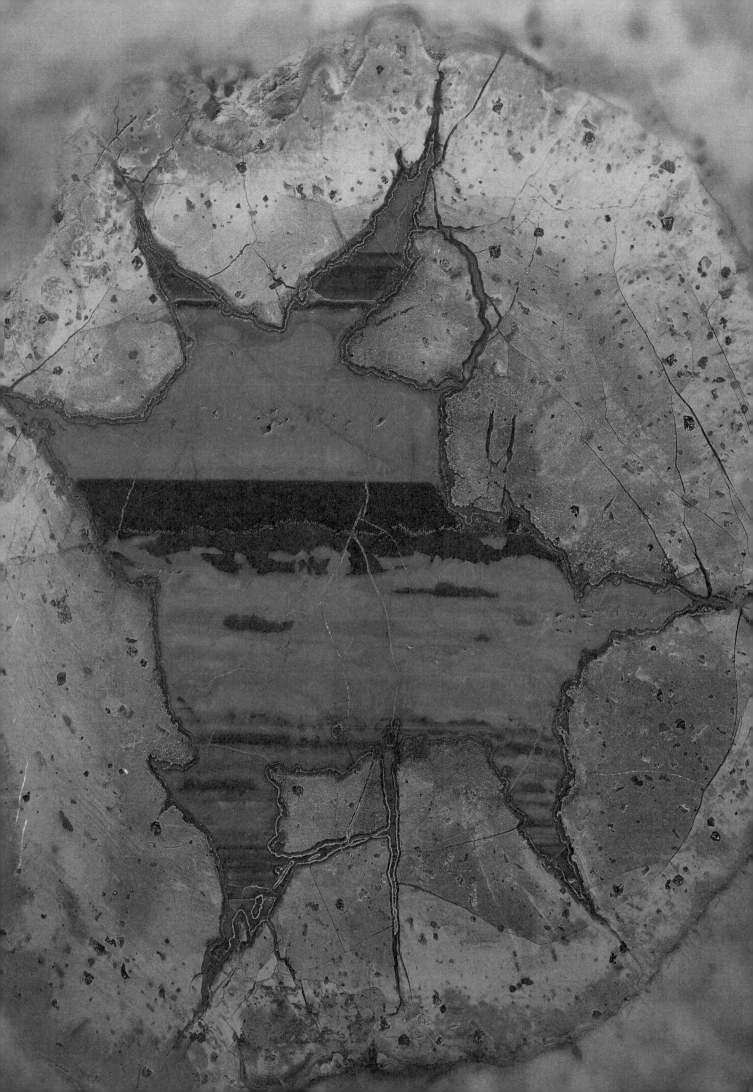

Agate is a hybrid term, widely used in the world of semi precious stones. The name goes back to Roman times and is first mentioned by Pliny the Elder, who refers to a then existing river Achate in Sicily, in which this type of stone was found. In a broader sense agate designates a type of stone, composed of 'chalcedony' (SiO_2), which is deposited, from aqueous solutions, inside a vein or hollow space in a rock, as described in the previous chapter.

Agates are known in many places around the world, most commonly in regions of volcanic rock. Very hard and resistant to weathering, pebbles of agate can be found far from their place of origin, in downstream river deposits or on pebbly beaches. The circumstances under which agates grow can vary from one place to the other, leading to great diversity in form, banding and color. Associated minerals, like manganese or iron oxide, often crystallize in 'dendritic' (branch-shaped) or rosette-shaped patterns inside the chalcedony. Veteran collectors can recognize the region, sometimes even the exact place of origin of an agate. Agate has been used for decoration and as a semi precious stone by many cultures for thousands of years. Large pieces may be worked into such objects as bowls or vases, smaller ones into animal figures or other trinkets. Very fine qualities are used for jewelry. Most striking are cameos and intaglios, those delicate gems for brooches or necklaces that are carved in such a manner that differently colored layers in the agate accentuate the renderings. The plates in this chapter show polished surfaces in reflected light, as well as translucent slices photographed in transmitted light. The fantastic images that materialize in agates defy the imagination.

SUB MARE This fine detail of a polished 'moss-agate', not much larger than a thumbnail, looks like sea-grass, gently moving with the heaving of the waves. But actually these are dendritic crystals of manganese oxide, not really organic at all. Here again, one feels that the separation between these two domains, the organic and the inorganic, does not have to be taken as absolute as is always implied.

ESCARPMENT Delicate dendrites and rosettes of a manganese oxide adorn a milky white agate, in a way suggestive of a sub-aquatic scene in a coral reef. This thin, polished slice measures 16 x 30 mm. and is photographed with transient light.

UNDER THE VOLCANO Thin bands of iron oxide inside translucent chalcedony bend upwards in the same way that huge strata curve up on the slope of a volcano. The title of this exposure was inspired by Malcolm Lowry's superb novel, in which such an ominous suspense burns under the surface of the narrative. Actual size 50 x 70 mm.

INDIAN SUMMER The intricate finesses of dendritic mineral growth, in this landscape agate, are a challenge to anybody's imagination. Actual size 13 x 22 mm.

BAROCK A garland of white chalcedony decorously girdles the periphery of this slice of agate that comes from the renowned area around Idar-Oberstein, referred to in the introduction of the previous chapter. Actual size 45 x 60 mm.

PURGATORY Through fire and torments the heavenly portal beckons'. Somehow a notion of the purgatory radiates from this thin slice of agate. Actual size 50 x 70 mm.

LITTLE SHIPROCK Shiprock is a conspicuous, pointed mountain of dark basalt, protruding 500 meters out of the desert plane in New Mexico, USA (see page 19). The precipitation of manganese-oxide, in this nail-sized landscape agate, has an allusive resemblance to the real Shiprock, including the crested dykes that fan out from the central mount into the desert.

TORRES D'ESPUMOSO This and several other pictures in this chapter are made of agates from a location called Espumoso, in Brazil. These sophisticated 'landscape agates', speckled with refined deposits of manganese- and iron-oxide, are much sought after by collectors. They are cut and polished into thin little slices, an art mastered by the best professionals of the trade only. Actual size 18 x 25 mm.

IM RIESENGEBIRGE Morgen im Riesengebirge is an allegorical painting by Caspar David Friedrich, depicting a crucifix planted on a ridge of jagged rocks in the Riesengebirge of eastern Germany. This image was created millions of years earlier, by the whims of nature, in a small slice of Brazilian landscape agate . It might well have astounded the artist. Actual size 20 x 25 mm.

BOTSWANA AGATE Botswana agates are very special,

characterized as they are by their reddish and blue hues and striking parallel banding. Small and delicate as they can be, they are often applied as cabochons or other trinkets in jewelry. Actual size 30 x 40 mm.

BLACK BEAUTY The slender profile of a Hindu dancer could be perceived in the elegantly curved bands of light and dark chalcedony in this agate from India. Actual size 40 x 50 mm.

THE ADORATION The images created by the precipitation of chalcedony and quartz inside geodes are infinitely imaginative. In this slice, about the size of a prayer card, one might see a rendering of the adoration of the Holy Virgin.

DENDRITES Dendrite crystals look like trees or branches. They are usually iron or manganese oxide that has precipitated from aqueous solutions inside narrow fractures in a rock, in this case inside chalcedony. The 'trees' in this winter landscape are about 7 mm. tall.

CAPTAIN NEMO'S RETURN Jules Verne's infamous hero and his ill-fated crew may well have perceived scenes like this when their submarine, the 'Nautilus', ascended to the surface by way of a secretive cave, in between one of those prolonged journeys in the deeps of the ocean. The picture is taken with transmitted light through a slice of agate, cut from a geode with an actual width of about 120 mm.

PTEROSAUR Could that disputed Mesozoic creature, half-dinosaur, half-bird, have looked somewhat like this effigy, diving headlong for it's prey? This is a sliver of manganese oxide inside a matrix of chalcedony. Actual size 45 x 60 mm.

GENESIS This strange and alien figure was created inside milky white chalcedony by theprecipitation of manganese oxide. Actual size 40 x 50 mm.

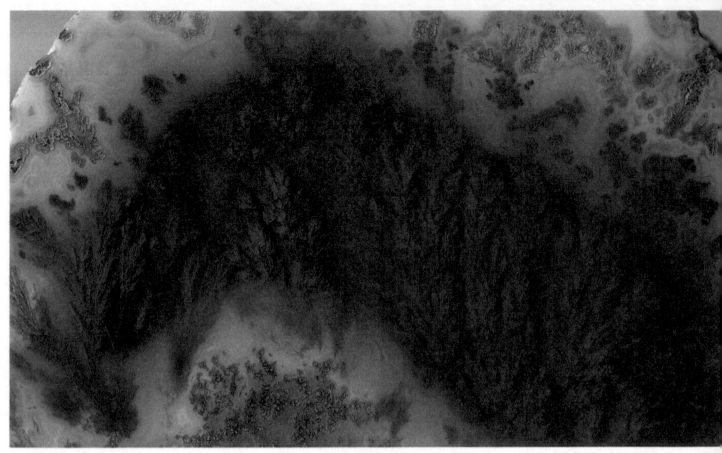

SUB MARE

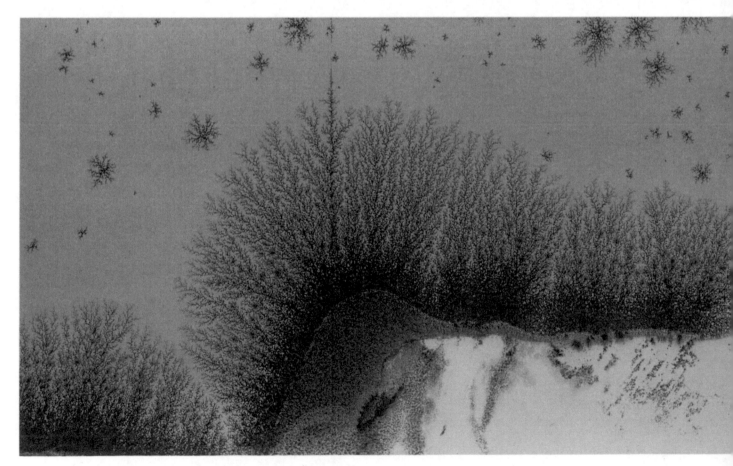

ESCARPMENT

UNDER THE VOLCANO

INDIAN SUMMER

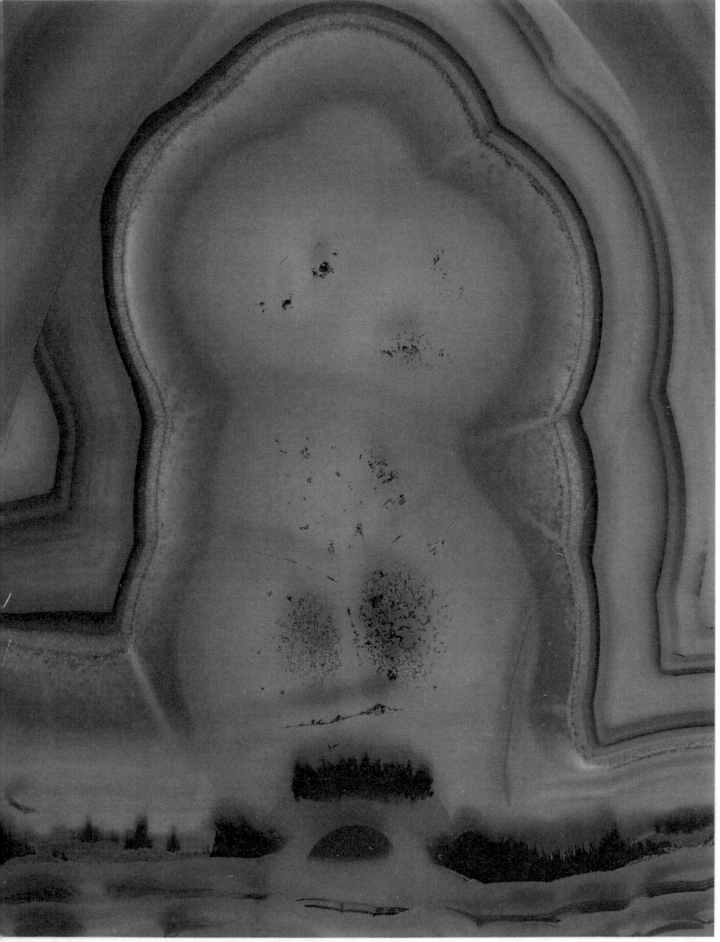

LITTLE SHIPROCK

TORRES DE ESPOMOSO >

IM RIESENGEBIRGE

< BOTSWANA AGATE

BLACK BEAUTY

< THE ADORATION

DENDRITES

CAPTAIN NEMO'S RETURN

PTEROSAUR   GENESIS >

Gem is a rather indeterminate description for a wide range of precious or semiprecious stones and minerals that are cut and polished for use in jewelry or as ornaments and trinkets. I this chapter the term is used, somewhat arbitrarily, to designate various minerals and stone types that are cut and polished primarily to bring out the beauty and imagery of the stone as such; these are not all gems that are worked into jewelry. They include jasper, a rock type much akin to agate, as well as malachite and opal. Of particular interest in this series is 'Pietra Paesina' from Tuscany in Italy, of which several remarkable examples are presented in this chapter. For an explanation see 'Pietra Paesina' hereunder.

PAESINA HARBOUR The artistry in this polished slab of Paesina Stone could be mistaken for a detail of a marine painting from the 17th century Dutch school. Actual size 130 x 160 mm.
The nature of Paesina Stone is explained in detail under Pietra Paesina, see below.

TERRA DI RIMAGGIO This sweeping landscape is laid out in a little slab of Paesina stone from Tuscany. Actual size 80 x 100 mm.

PAESINA TEPUI Tepuis are the notorious, flat-topped mountains located in the savanna region of southern Venezuela. The slopes of these high and isolated mountains are so steep and inaccessible that each has a particular fauna and flora, often including unique species found nowhere else. The tepui here depicted is an allegory, seen in a small, polished slab of Paesina stone from the rolling hills of Tuscany. Actual size 90 x 120 mm.

VERDE D'ARNO Pebbles of this type of rock, which is akin to Paesina Stone, are found in the valley of the Arno River in Tuscany, Italy. Although probably formed by similar processes to those that created the patterns in Pietra Paesina (see below), the structures in Verde d'Arno are more geometric. These could have inspired painters like Lionel Feininger or Paul Klee. Actual size of this polished face is 70 x 90 mm.

PIETRA PAESINA Pietra Paesina, also known as 'ruin-marble', is an amazing type of stone found in the Tuscan hills around Florence. The ancients already considered it a material worthy of a craftsman. By origin this rock was a lake- or river-deposit-ed sediment of finely alternating, calcareous silts and clays. Subsequent deformation and diagenetic processes introduced little cracks and displacements in the rock as well as dendritic precipitation of manganese. Even electro-chemical processes have been proposed in an attempt to explain the complex structures in this amazing rock. Certainly, the forces that acted must have been as subtle and original as the refined artistry found in this stone. Other plates in this chapter show some more examples of Paesina Stone. Actual width 120 mm.

SONGLINES Songlines are the spiritual routes of travel and dance across Australia as traversed and worshipped by the aboriginal people of that continent. In this fist-sized, polished chunk of jasper from Roy Hill in Western Australia (see next plate) one might see a microcosmic version of the songlines.

ROY HILL ENCOUNTERS This type of jasper is found in the vicinity of Roy Hill, a remote post in the outback of Western Australia. In the course of geologic time this type of rock has been partially stained red by iron.
The remaining, unstained patches often take on enigmatic shapes, like the 'dramatis personae' in the polished slabs here juxtaposed. Actual size 100 x 120 mm.

SPECTROLITE Spectrolite, also called labradorite, is a variety of the extensive group of rock-forming minerals called feldspars. Large, tabular crystals often display iridescence, a changeable and magic emanation of colors resulting from the interference of light reflected by the semi-translucent, lamellar crystal planes. Actual size 40 x 50 mm.

CORDIERITE Cordierite is a characteristic mineral formed by alteration in highly metamorphic rocks. It is akin to beryl, and clear crystals of appropriate color are sometimes used in jewelry. The crystals are often transected by fissures, along which replacement by other minerals takes place. That feature is called 'pinitization' and it is well displayed in this thin section.

OPAL Opal is hydrated, amorphous silica, deposited from silica-bearing waters in voids and fissures of certain rock formations. Water molecules trapped in the silica account for the milky luster and brilliant pastel colors that make opal into such a desirable precious stone. Much of the high-quality opal traded these days comes from Australia. A number of years ago this continent officially declared opal to be their national gem. This cut and polished Australian specimen is a Yowah Nut, a type in which the opal is intercalated in the concentric fissures of a limonite concretion. Actual size 30 x 50 mm.

MALACHITE Malachite, a copper carbonate with vivid green colors and banded, kidney-like structures, is hard to mistake. This polished slab comes from the Congolese copper-mining province Shaba. Malachite, although closely associated with the copper ore, is not an ore in its own right. Often malachite is worked into trinkets and other small ornaments. More lavish was its use to adorn the interior of tsarist palaces in Moscow and St. Petersburg. This extravagance before long led to the total depletion of the renowned malachite mines in Russia. Actual size 90 x 110 mm.

RUBY IN ZOISITE Longida, in Tanzania, is a unique locality for the occurrence of outsized, elongated crystals of ruby, embedded in a green matrix of zoisite. This picture is of a thin slice, craftily sawed off perpendicular to the long axis of the crystal and polished on both sides. It is photographed with transmitted light, so that all the intricacies of the ruby's crystal structure can be well observed, including growth planes, cracks, and fissures that resulted from stresses during crystal growth. Actual diameter 100 mm.

BLUE STALACTITES Milky blue chalcedony has here been deposited in the form of small pinnacles and a sugary crust. As explained for geodes and agates in the previous chapters, most chalcedony is formed inside rock cavities, often in a rich variety of form and color. The pinnacles in this cavity are actually small stalactites, formed by the drip-drip of water inside a small cavity, just like the formation of the much larger, calcitic stalactites in caves. Actual size 60 x 80 mm.

THE FLYING DUTCHMAN The ragged and ghostly ship of the legendary Dutchman, who was condemned to an eternal quest of the world's seas, may have appeared somewhat like this to the bewildered seamen that ran across him, navigating their ship across the waves, during faraway, stormy nights.
This scene occurs in a polished slab of Paesina Stone, see explanation under Pietra Paesina. Actual size 100 x 130 mm.

LUCIFER The mythological bearer of light materializes inside

a rhyolitic agate from West Australia. Rhyolite is a characteristic type of extrusive rock (see page 10) that is often the host rock of Australian geodes. Actual size 35 x 45 mm.

PAESINA HARBOUR

PAESINA TEPUI >

TERRA DI RIMAGGIO

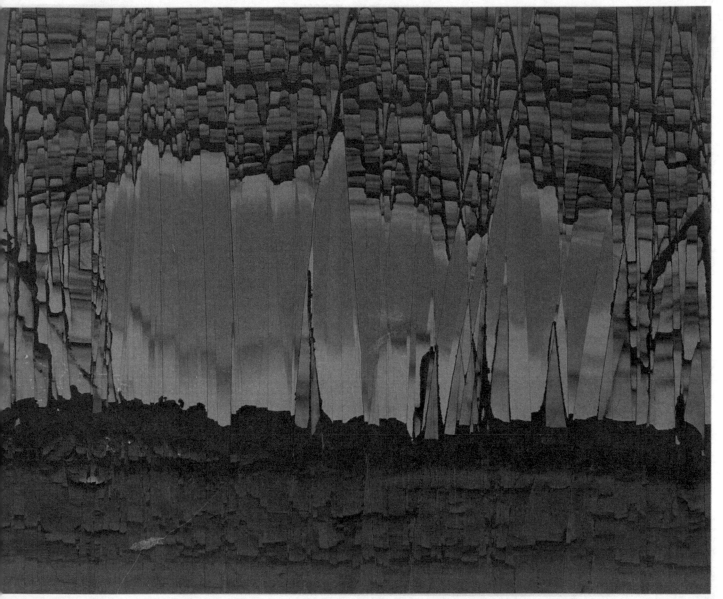

< VERDE D'ARNO

PIETRA PAESINA

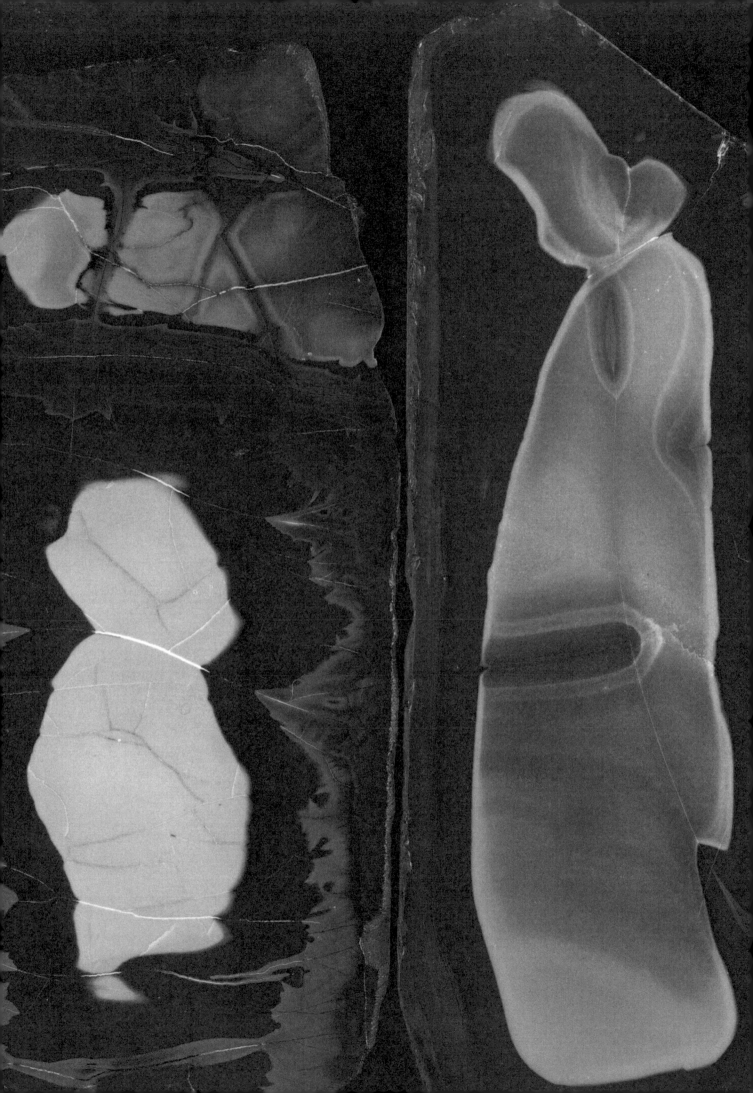

SPECTROLITE

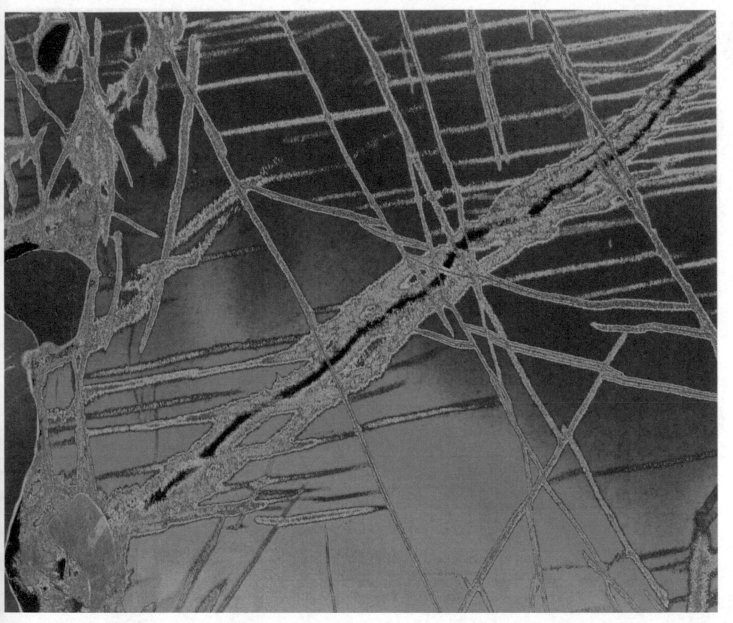

CORDIERITE

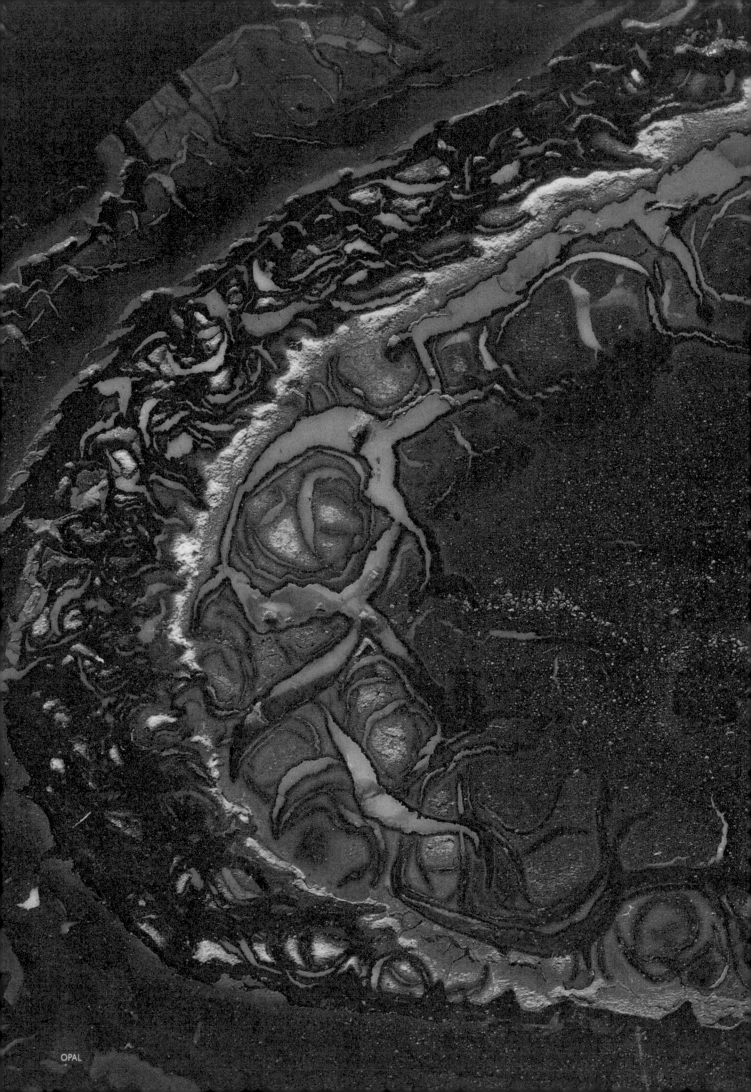

OPAL

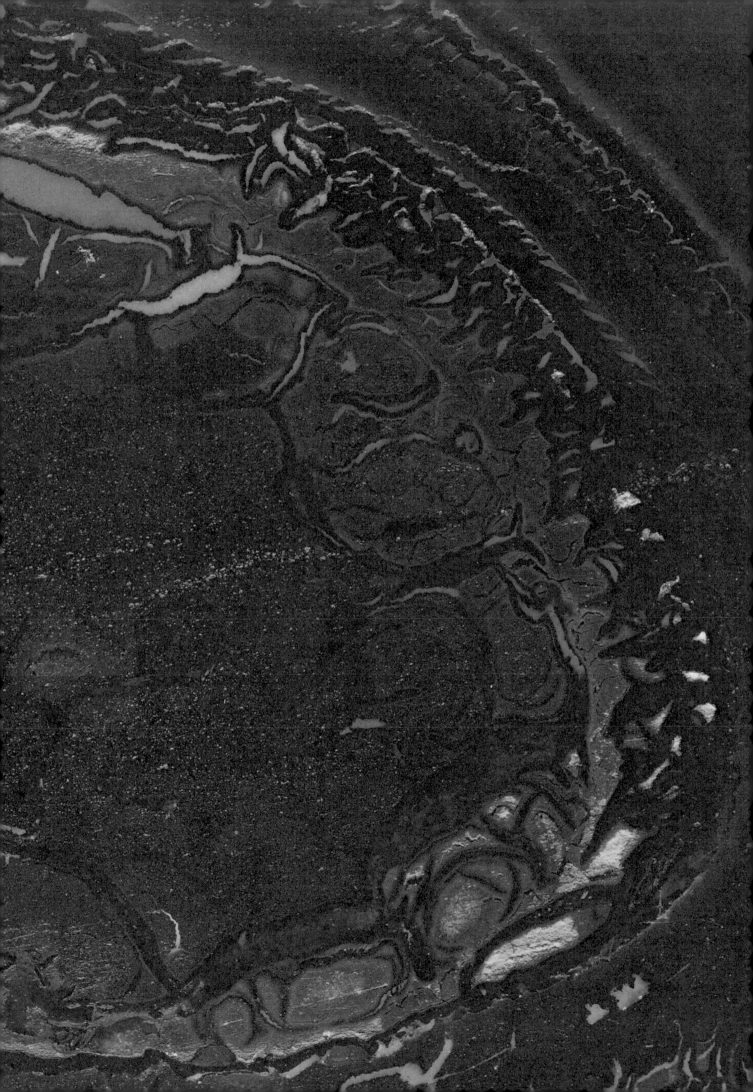

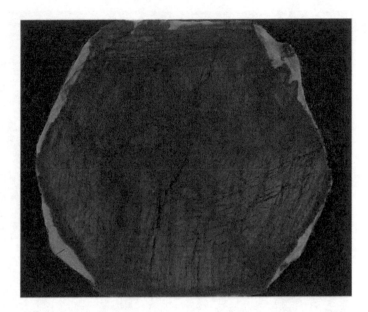

< MALACHITE

RUBY IN ZOISITE

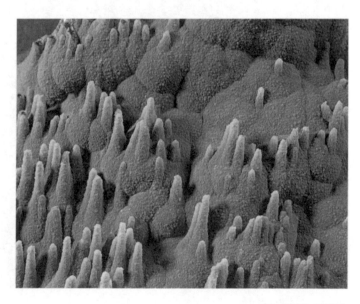

BLUE STALACTITES

THE FLYING DUTCHMAN

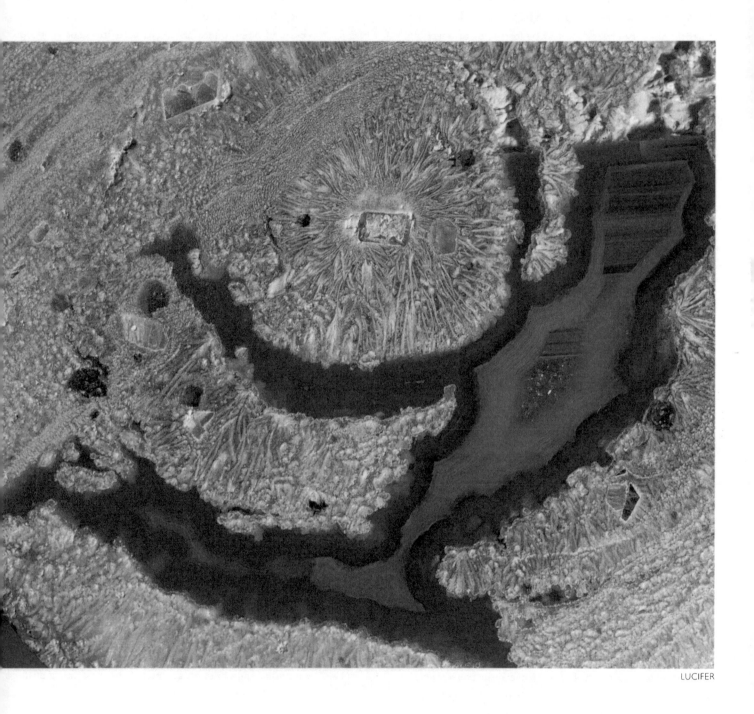

LUCIFER

The word 'texture' refers to the detailed structure and makeup of all kinds of materials, such as textile, rocks and many others. The pictures in this chapter illustrate that textures in minerals and rocks are very diverse and may be formed by a variety of processes, like crystallization and sedimentation, but also through pressure, deformation or geochemical processes. Many pictures in this paragraph are of 'thin sections', wafer-thin (10 to 40 micron) slices of rock, which are glued on to glass and prepared with special skill. Thin sections are one of the fundamental means for analyzing rocks and minerals, and geologists routinely examine them with a polarizing microscope. The inspection is carried out with the help of various optical means and with transmitted light that may or may not be polarized. When the light is polarized, crystals often take on striking colors, due to birefringence of the light beams.
The actual sizes of the thin sections shown in this paragraph and the next are in the range 10 to 50 mm. and not mentioned separately in the texts.

CARNALLITE Carnallite is a rock salt and as such belongs to a group of sediments referred to as evaporites. Evaporites form part of marine sedimentary sequences and are precipitated by the drying up of isolated parts of continental seas during dry and warm climates. A present day example is the Dead Sea of Israel. In thin section and polarized light, the tightly interlocked salt crystals take on attractive colors.

BISCHOFITE Bischofite is a magnesium containing rock salt, like the carnallite of the previous plate. These two pictures are of salts mined, by slurry pumping, from the deeper subsurface in the north of the Netherlands. The thin sections here photographed are rather exceptional, since a special technique is required for their preparation and preservation.

BORDERLINE The fine, darkish material in this thin section is basalt, a magma that has solidified at, or near, the earth's surface. When ascending to the surface from the deep underground, such magmas may bring along bits and pieces of other rocks. Some of these pieces do not melt, such as the chunk of peridotite on the right hand side of the picture, in which the olivine crystals display their characteristic, lively colors of interference.

MUSCOVITES Muscovite is a micaceous mineral occurring frequently in plutonic and metamorphic rocks. It is the mineral that often gives the sparkle to granite and gneiss or appears as tiny stacks of wafers in other plutonic rocks (see Aquamarin, page 54). In thin section and polarized light, micas display fading colors of birefringence and a characteristic 'goose-flesh' texture.

PERIDOTITE Peridotite is a 'plutonic' rock. It is formed at great depth, where the earth's crust changes into the semi-solid mantle. It is assumed that most of the earth's mass, notably the mantle, has a peridotitic composition, but this has never been proven (the earth's interior is simply too deep and too hot for sampling, we know less of it than we do of the surface of the Moon or Mars). Many meteorites, i.e. those of the stony variety (the other variety is metallic), also have the same composition, which suggests that peridotite is a key building material of the whole solar system. The colorful olivine that makes up the bulk of peridotite, has been partly serpentinized, which means that it has been altered, along the grain boundaries, into the water rich mineral serpentine. This accounts for that craquelée look of an old oil painting.

TIGER EYE Tiger-eye is a fibrous mineral, belonging to the silicate group. It is well known for it's silky, golden gloss and is often cut and polished into cabochons and trinkets. Here, in thin section, its appearance is more that of a sanguineous chunk of meat.

MYLONITE MOSAIC Shown here in thin section is mylonite, a thinly laminated, metamorphic rock. Fine-textured seams of mica are intercalated with thin bands of quartz. The quartz grains have undergone recrystallization and have thus been rearranged into a close-fitting mosaic.

HORNBLENDE Hornblende belongs to a group of minerals called the amphiboles, which occur in igneous as well as metamorphic rocks. In this thin section exposure the delicate texture inside the crystals is accentuated by the subtle gradations in the colors of birefringence.

PERTHITE Perthite, sometimes incorrectly referred to as a feldspar mineral, is actually a texture of fine, interspersedlayers occurring in alkali feldspars. This lamellar texture results from a separation of sodium- and potash-feldspars during cooling and solidification of the mineral. Thin section, polarized light.

CHIASTOLITE Chiastolite is a unique variety of the mineral Andalusite, an aluminum silicate. Chiastolites are columnar, prismatic crystals of this mineral, with dark inclusions of carbon or clay particles, which are arranged in the form of a cross. Chiasma is Greek for cross. This picture is of a thin section cut squarely across such an elongated crystal column. Diameter ± 45 mm.

GALENITE Much like contour lines on a map, crystal-growth patterns (see 'Striations', hereunder) accentuate the surface of a small specimen of galenite, a common lead-ore. Actual size 40 x 50 mm.

LACQUER PROFILE Seen here in cross section are beds of sand and gravel deposited in a Holocene fluviatile environment in the Netherlands. Cross bedding, ripple marks and other structures typical for such river deposits are very distinctly displayed. This section is characteristic for the way geology is practised in the predominantly young and unlithified sediments of Holland. It is a 'lacquer profile', prepared by immersing a clean cut, vertical section of such sediments, for example in a sand- or gravel-pit, with lacquer. After the lacquer has hardened the section is carefully peeled off and fastened to a board. The width of this section is nearly one meter.

STRIATIONS Striations and other striped patterns sometimes seen on crystal planes are a telling indication of how crystals grow, layer-by-layer. See also previous page. This example is on the face of a pyrite-cube. Actual size 17 x 22 mm.

GRAPHIC GRANITE Granite is formed at great depth, by the solidification of magma. Graphic granite is a particular variety, in which the two essential rock forming minerals, quartz and feldspar, have solidified eutectically (simultaneously). This results in an orderly pattern: the quartz takes on the shape of hieroglyphs and is embedded in a matrix of patchy, monocrystalline feldspar.

STRICKBLEIGLANZ This cut and polished slab of lead ore, typical for certain ancient mining districts in Germany and Belgium, shows a remarkable skeletal pattern, which results from the intergrowth of zinc in the lead ore. The Germans refer to this pattern as 'gestrickt' (knitted), hence the striking name they give to this type of ore. Actual size 100 x 150 mm.

BORDERLINE   MUSCOVITES >

PERIDOTITE

TIGER EYE

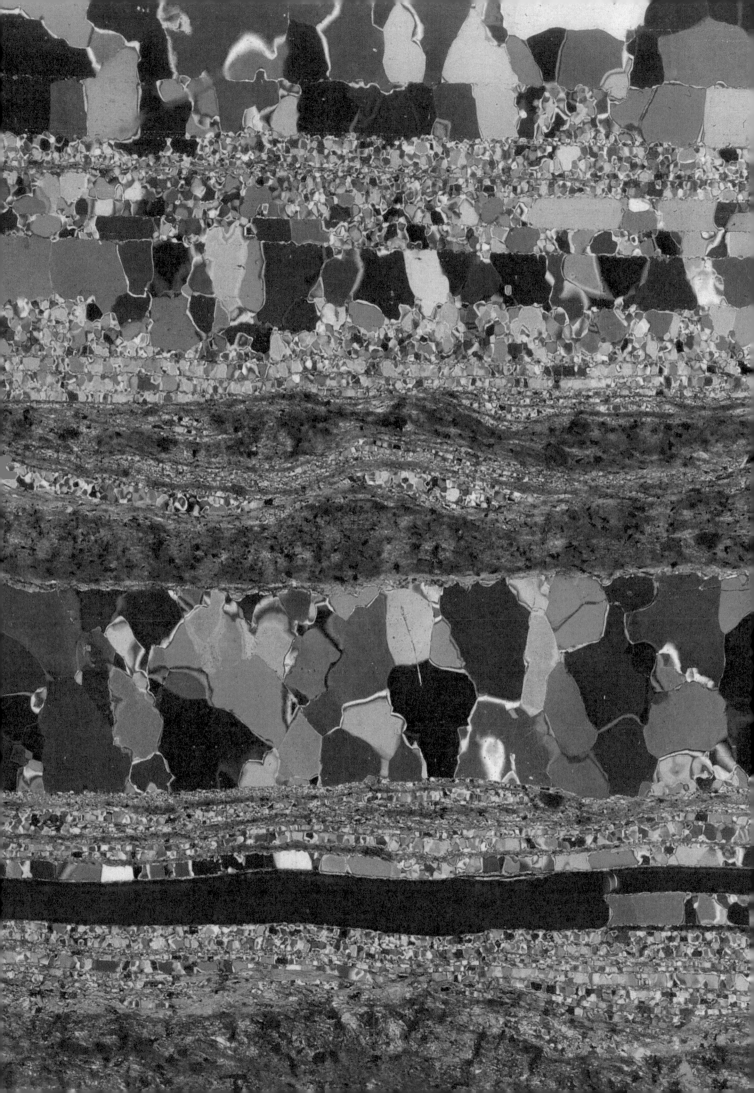

HORNBLENDE

CHIASTOLITE >

PERTHITE

< GALENITE

LACQUER PROFILE

STRIATIONS

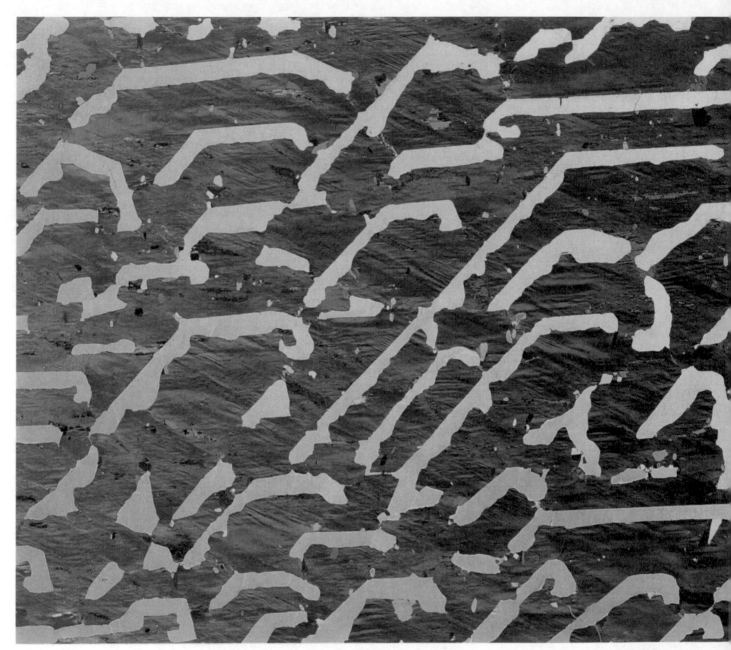

GRAPHIC GRANITE

STRICKBLEIGLANZ

In the geological record, deformation plays a key role: on a large scale, that of whole mountain ranges, but also in tiny portions of minerals and rocks. The principal drivers are the continuous mobility and flow of the hot and viscous material in the earth's core and mantle, which are conveyed to the 'lithosphere', i.e. the rock formations of the earth's outer crust.

The implications of these phenomena are better known since the elaboration, in recent decades, of the notions of sea-floor spreading and continental drift, which are unified under the concept of plate tectonics. According to this now widely accepted model, once upon a time there was one large, coherent mass of continental crust, sometimes referred to as 'Pangaea'. This single mass of lithosphere started to break apart into several large chunks some 250 million years ago. These chunks, now known to us as the individual continents, move about on the hotter and more viscous mantle-material beneath, much like a film of grease floating on a plate of soup. No wonder that this leads to strains, stresses, and deformation in the rocks that form the crust. Most of this mobility occurs along the coastal margins of the continents. These are thought to be the zones where thin, oceanic crust is pushed down under the thicker, continental crust. But continental plates also collide with each other, like in some parts of the near East and southern Asia. Earthquakes and volcanism, the most telling evidence of all this mobility, occur most frequently in these zones, for example in the Circum-Pacific realm, in Turkey and in Greece.

'Panta Rhei' (everything moves) said the old Greeks and with first hand evidence they knew this to be true not only for the sea around them, but also for the ground beneath their feet.

MICROFOLDS Originally these were thin, sedimentary layers of quartz sand and clay with plant remains, probably deposited in a riverbed. But in geology nothing is static, everything is transformed and deformed: minerals, rocks, the face of the earth itself. Slow, but massive, movements dragged these sediments to the deep subsurface. The sand and clay beds were altered into carbon-rich schist and quartzite, and intense compressive forces warped the thin bands into steep microfolds. Eventually, uplift and erosion brought it all back to the surface. This example is small-scale (thin section, height 20 mm.), but folding of bedded rocks occurs on all scales, up to that of whole mountains.

FOLDED BIF BIF, short for Banded Iron Formation, is a characteristic type of rock composed of thin, alternating beds of jasper, hematite and tiger-eye, which may, to greater or lesser degrees, be deformed into folds or fragments. This type of rock, of which polished surfaces are pictured here and in several of the following plates, is the main iron-ore in some of the world's largest mining districts, most notably in West Australia. Many BIF's are very ancient sediments, some as old as 3 billion years and particularly interesting in that they represent a crucial period of evolution on our globe. They originally were sediments deposited in shallow seas, where primitive bacteria brought about the oxidation and precipitation of the iron dissolved in the seawater. This led to the recurring deposition of thin beds, probably under the influence of climatic cycles. This was a new phenomenon on planet earth and an indication of crucial changes in the earth's evolution, notably in the atmosphere. Massive sediments were thus accumulated and, after a long history of deep burial and exposure to tectonic movement, these rocks were intensely altered and deformed. This polished slab is a telling example of such deformation on a small scale. Actual size 180 x 220 mm.

CLEAVAGED VEIN Intense tectonic forces, that imparted a cleavage type folding (see last picture) to a schistose rock, have also buckled up a thin stringer of quartzite transecting it. One might well imagine to be looking at an arctic landscape, with a river of ice meandering across the dunes of a barren desert. Thin section.

KINKS Kinks are a characteristic type of microstructure, also called zigzag folds or knee-folds. They are straight-limbed, angular folds with narrow hinge zones that may occur in thinly bedded rocks or in crystals with a laminated structure, like in the feldspar displayed here. Kinks form in response to intense, compressive forces, with slip movements occurring along the laminar plane surfaces. Thin section.

MINIGRABEN In geology, a graben denotes a large structural feature, a chunk of the earth's crust that has sunk downwards along two sub-parallel faults. The well-known Rift Valley of East Africa marks the site of a graben, as do parts of the Rhine Valley and other marked depressions in the globe's surface. In cross section a graben would look much like what is seen in this picture. In fact the miniature graben formed in this polished slab of

BIF (see under Folded BIF) is more or less a result of the same forces and conditions that are at work in forming the large-scale grabens. Actual size 25 x 40 mm.

MARAMAMBA In this polished slab of BIF, the soft greens and reds of jasper, the metallic glow of hematite, and the golden fibers of tiger-eye evoke the likeness of an alien landscape, remote and otherworldly. Actual size 90 x 120 mm.

APEX In this polished slab of BIF, from Western Australia, a thin seam of hematite has been deformed and fragmented into a crested form by the same stresses that generated the fibrous texture of the surrounding tiger-eye. Actual size 75 x 100 mm.

BIOTITE IN ORTHOPYROXENE A flake of biotite, a micaceous mineral, has been buckled up, like a pennant in the wind, by internal stress forces inside a holocrystalline mass of orthopyroxene. The image is of a thin section of a rock sample from the Bushveld complex, South Africa. The colors result from polarization of the transient light used to make the photograph.

SNOWBALL GARNET This thin section demonstrates the rotational deformation of a rigid object, in this case a large garnet crystal, that is suspended in a fine grained, more readily deforming rock. Thin section, polarized light.

PRESSURE FRINGE Pressure fringes, also called pressure shadows, are a type of microstructure formed by intense shear forces in deep-seated rocks. This example shows a darkish brown spherule of pyrite in a schistose host-rock. The pyritic spherule rolled along but remained rigid while the schist was deformed. Quartz, formed in the 'wake' of the rotating spherule, was stretched out into a fibrous and curved shape that looks much like an old-fashioned airplane propeller. This thin section is prepared from Archaean rock in the Yilgarn region, Western Australia, which is some 2.5 billion years old!

WAVY CALCITE This thin section shows a detail of a calcite vein precipitated inside a fracture in a rock, from hot fluids, deep in the earth. Deformational stress during the precipitation of the calcite brought about the curved, fibrous shape of the crystals.

CRENULATION CLEAVAGE 'Cleavage' is an advanced stage of microstructural faulting and deformation in a rock, whereby the rock in question often takes on a micaceous, finely foliated fabric. Crenulation cleavage, as seen in this thin section, results from successive phases of such deformation.

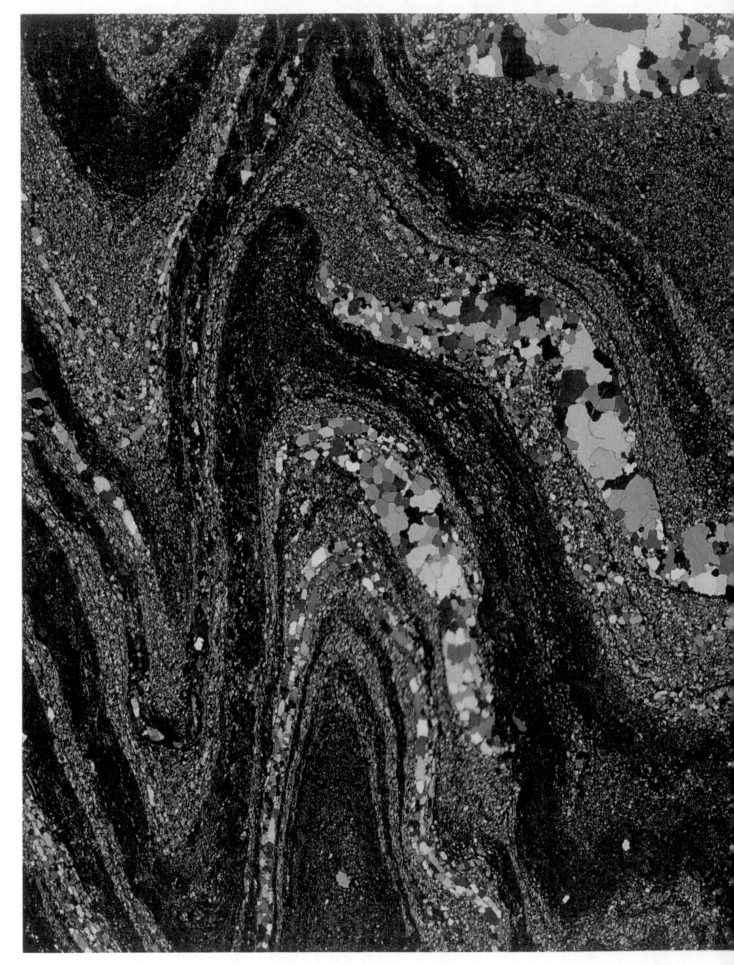

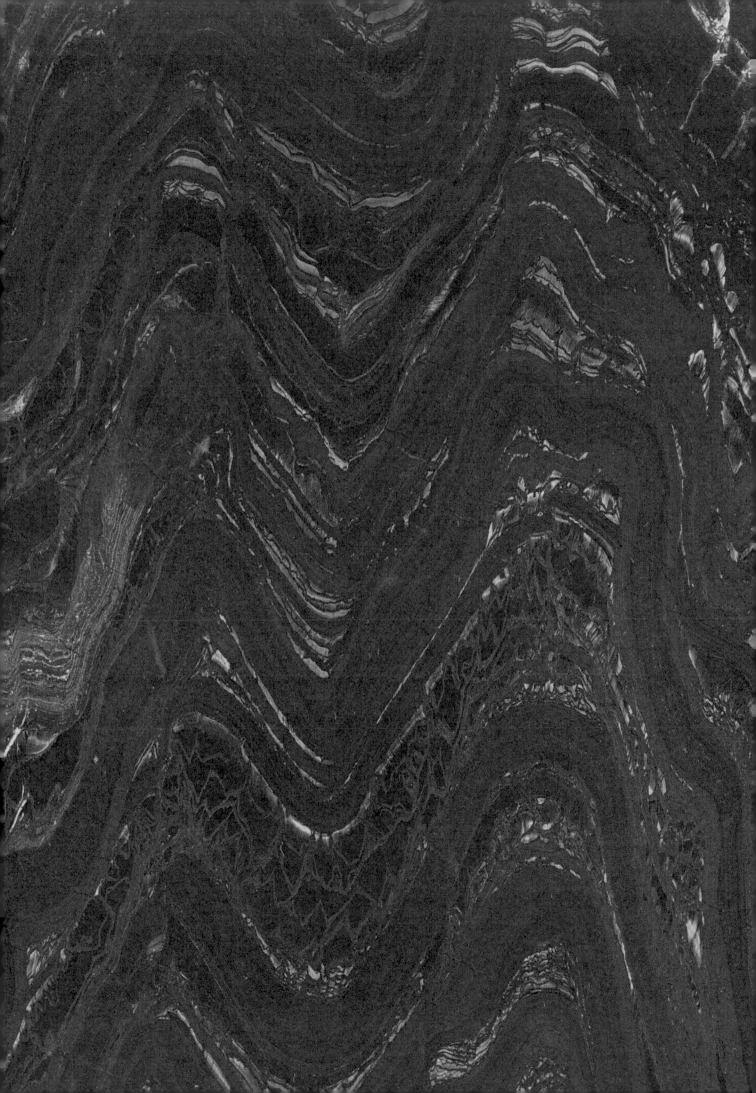

CLEAVAGED VEIN

KINKS >

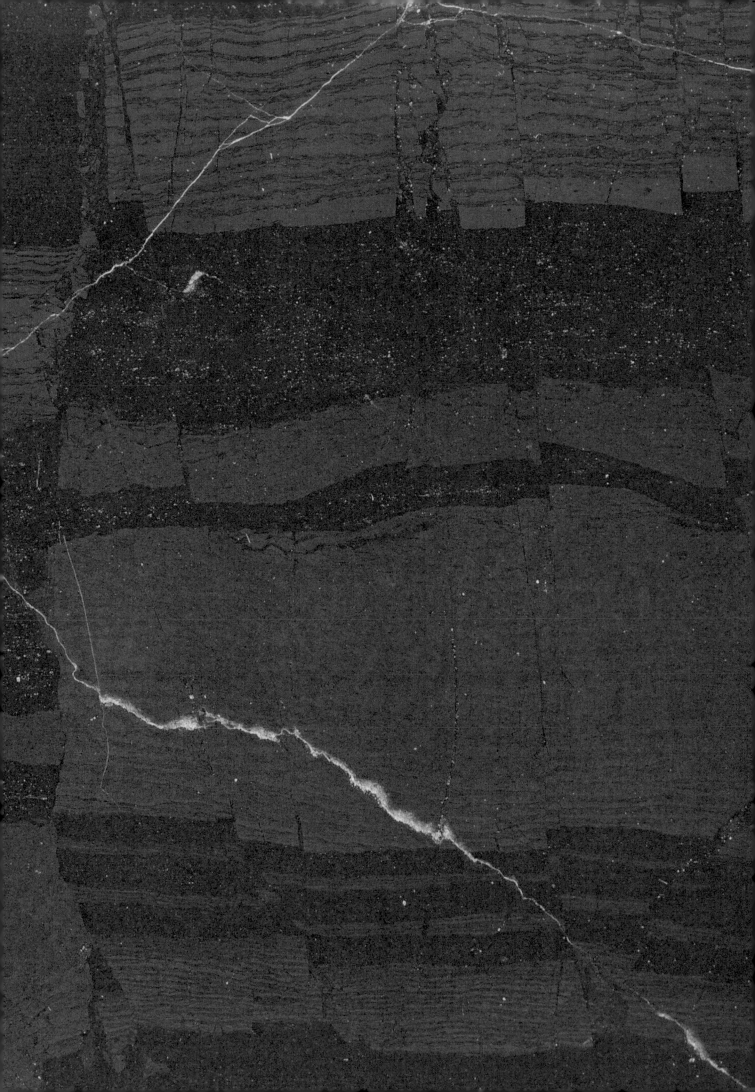

MARAMAMBA

APEX >

BIOTITE IN ORTHOPYROXENE

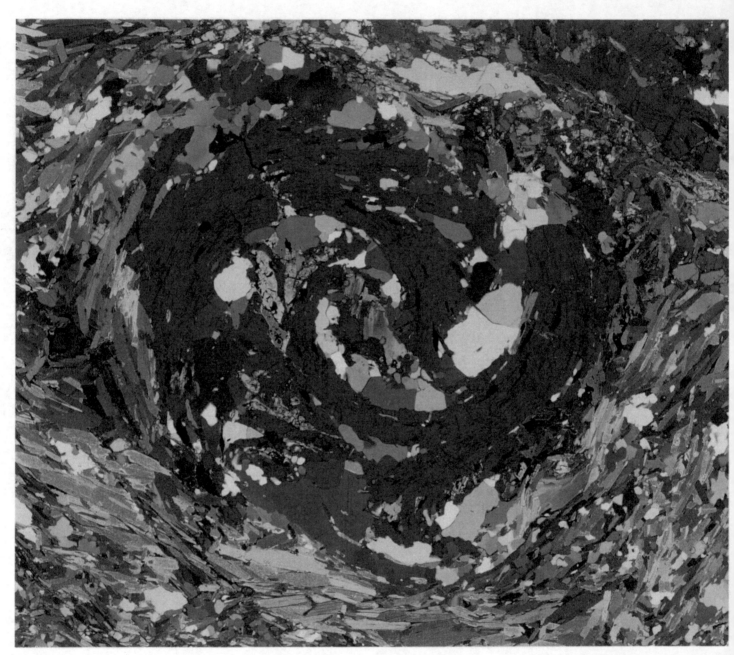

SNOWBALL GARNET

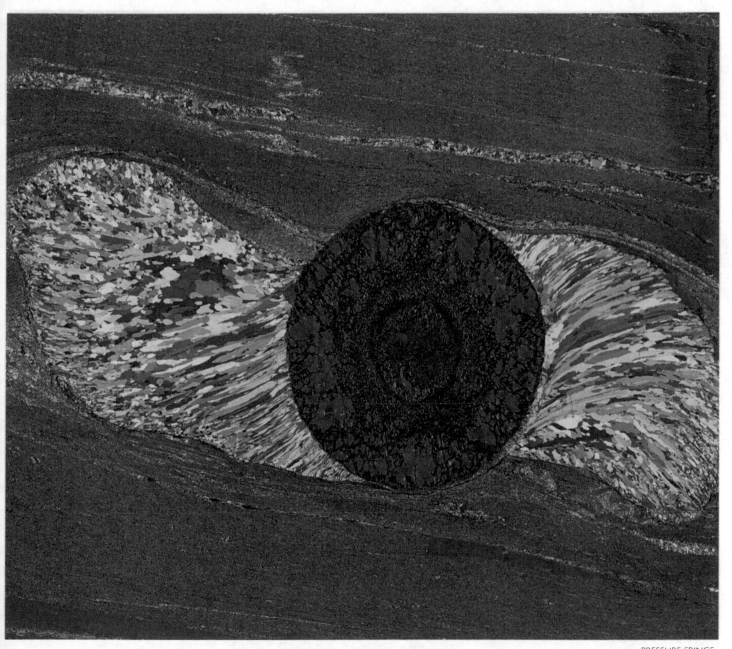

PRESSURE FRINGE

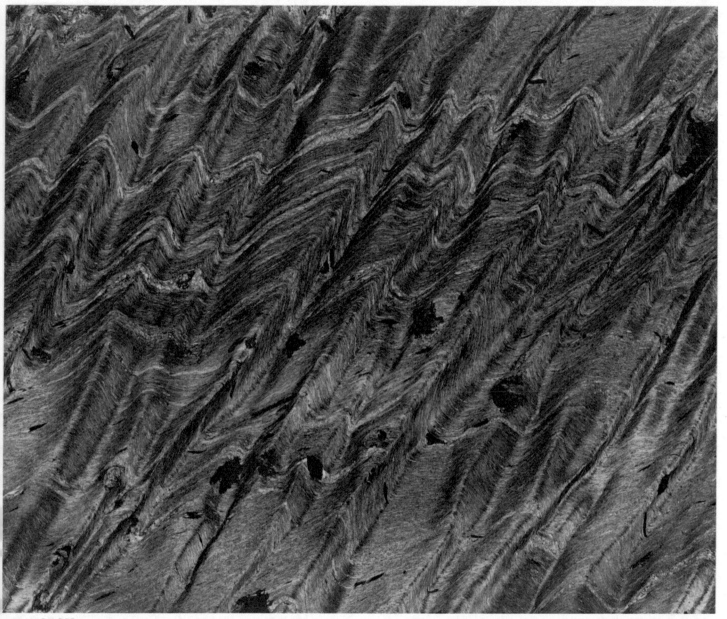

< WAVY CALCITE

CRENULATION CLEAVAGE

Fossils are the remains of organisms that lived on Earth in the geologic past. The fossil record ranges from the most primitive, unicellular life forms of three billion years ago to the highly sophisticated biota of recent times. Such remains are usually petrified, but some have remained organic, like plant remains in lignite and coal, insects in amber or a frozen mammoth in polar ice. Fossils are the underpinning of our knowledge of past life and the evolution of species. It is perhaps ironic, but thanks to petrifaction, selective weathering, and also to the fact that fossils can be cut, polished, and observed in thin section, certain attributes of an organism can be more readily observed than would have been possible in a live specimen! The images in this final chapter reflect Earth's life forms, what we call the organic. The previous chapters reflect the non-organic, or the lifeless forms. But if the reader should feel like turning back the pages (wondering whether he has not already seen this or that form), he may well question whether this differentiation is really that unambiguous.

LILIES OF THE SEA These menacing tentacles are nothing but the fossil remains of crinoids, tranquil sea lilies that gently swayed on the bottom of a Devonian sea, some 400 million years ago. See also next picture. Actual size 320 x 450 mm.

CRINOID BRECCIA Crinoids belong to a group of animals, the Echinoderms, which also include starfish and sea urchins. Contrary to their appearance and the nickname 'sea lilies', crinoids are animals and not plants. Crinoids have lived in shallow seas since the very ancient geologic past and they still exist today. They have a long, thin stem attached to the seafloor, topped with a flower-shaped body that bears feathery petals. The slab of limestone on which these fossil remains stand out so strikingly, comes from Devonian beds in Morocco. Actual size 600 x 750 mm.

FERN This lively cross section of the stalk of a fossil fern appeared on the face of a small pebble that was picked up on a beach in Tasmania and run through a tumbler. Actual width 30 mm.

LEPIDODENDRON Lepidodendron belongs to the Lycopsids, a class of plants that made up a large part of the lush vegetation in the extensive swamps of the Carboniferous era. The Lower Carboniferous is the more than ± 350 million year old geological period that accounts for most of the coal reserves on earth. This plate shows a fossil imprint of the diamond shaped pattern that is typical for the bark along the upper section of the trunk. These plants could achieve gigantic size, 30 to 40 meters and thus were larger than many trees. Descendents of the Lycopsids are still living today, but how heartrending to realize that it is only in the form of dwarflike mosses. Actual width 110 mm.

PETRIFIED WOOD Petrifaction, and later leaching, has accentuated the wood-texture much more dramatically than could have been seen in the live wood. Actual size 120 x 150 mm.

TRILOBITES Trilobites belong to an illustrious class of very old, extinct organisms, that lived from the early Cambrian, over 500 million years ago, well into the Permian period, some 250 million years ago. They were hard shelled, bottom-dwelling sea organisms, with some 15 000 species known today, ranging in size from 5 to 500 mm. With their tripartite, segmented bodies they belong to the phylum of Arthropods (same as spiders) and they were the first organisms on Earth outfitted with eyes. Fossils of trilobites are abundant and popular with collectors. The species here shown, measuring ± 25 mm., is Proetus Bohemicus from Devonian deposits in Morocco.

LAST DANCE OF THE FISH Although it looks as if these fossilized fish are performing an elegant water dance or a game of tag, it is more probable that they were in agony of death, struck by whatever calamity led to their sudden perishing. The species on this slab, hardly the size of sardines, are called Knightia and come from a well known fossil-hunting site in Wyoming, USA.

AMMONITE IN THIN SECTION Ammonites are cuttle fish with snail-like shells that lived it up in the Jurassic and Cretaceous seas, some 200 to 100 million years ago. The name relates to the ancient Sun God 'Ammon Re' and to the holy 'Ammons Horn' of the ram. Fossil ammonites are found in abundance in Jurassic and Cretaceous sediments all over the world, and have always been associated with fables and superstition. Actually proof exists that they were already used as talisman in the Stone Age. The many types varied in size from a few millimeters across to over one meter.

The insides of the tiny, fossilized specimen here depicted are filled with crystalline calcite and quartz. The picture is of a thin section, taken with polarized, transmitted light, which explains the lively colors taken on by the crystalline infill. Actual diameter 20 mm.

AMMONITE DECO Exposed here is the polished surface of a tiny ammonite, the size of a dollar, ingeniously cut in half. Its crystal filled chambers, and the gracious curves of the septa (partitions) in between, would certainly have appealed to the architects of the art-deco period. Diameter 22 mm.

SPHENODISCUS LENTICULARIS This fossil of the ammonite Sphenodiscus Lenticularis is a stone-core: an interior cast of this shellfish rather than the outer shell. Thanks to this the suture lines become much better visible. The suture lines are the edges of the partitions (septa) that divide the interior of the shell into separate chambers. These are few and far between in some ammonite species and perplexingly contorted and delicate in others, as in the specimen here seen. Actual diameter 60 mm.

AGATHICERAS TIMORENSIS This thin section of a tiny, fossilized ammonite has all the distinctions of antiquity. The creature lived somewhere in what is now the Indonesian Archipelago, some 150 million years ago, during the lower Cretaceous era. The specimen belongs to a large collection compiled, in the 1930's, on the Isle of Timor by Dutch paleontologists. Actual diameter 25 mm.

CAPTURED IN AMBER Some 40 million years ago, during the Eocene, this slender insect was trapped in resin, forever. Resin that spilled out profusely from pines and oaks in the forests then covering a region that is now the Bay of Gdansk and the Baltic coast of Poland.
With time, chemical processes transformed the resin into that lustrous, yellow to golden-red, semi-precious stone we call amber. Already in the Stone Age amber was collected and polished into valued jewelry. The ancient Greeks and Romans also obtained amber for their jewelry from these exceptionally rich deposits on the Baltic coast.

SPIDER LARVA This diaphanous larva of a spider has been incarcerated, since Eocene times, in amber from the Baltic coast of Poland. Fragments or whole specimen of the flora and fauna of that time, notably insects and spiders, were captured that way in resin, some 40 million years ago. Many fossils are in mint condition and provide valuable material for biologists and paleontologists who study that epoch and the evolution of species at large. Larger organisms, such as lizards and frogs, are only rarely found in amber, but they do occur and have tickled the fantasy of science fiction writers and romanticists alike.

THE MOSQUITO AND THE SPIDER Immortalized in a small morsel of amber, a mosquito and a little spider look like they are engaged in a ritual dance. Or rather a Danse macabre, since these insects were cruelly trapped in sticky resin, in an Eocene forest along the Baltic coast, more than 40 million years ago!

PETRIFIED TREE STUMP Victim of a volcanic outburst, some 65 million years ago, the fossilized trunk of this tree still proudly stands upright, on the very place where it grew. It overlooks what is now called the Bisti Wilderness (New Mexico, USA), a barren stretch of 'badland' where hardly anything will grow today!

LILIES OF THE SEA

CRINOID BRECCIA >

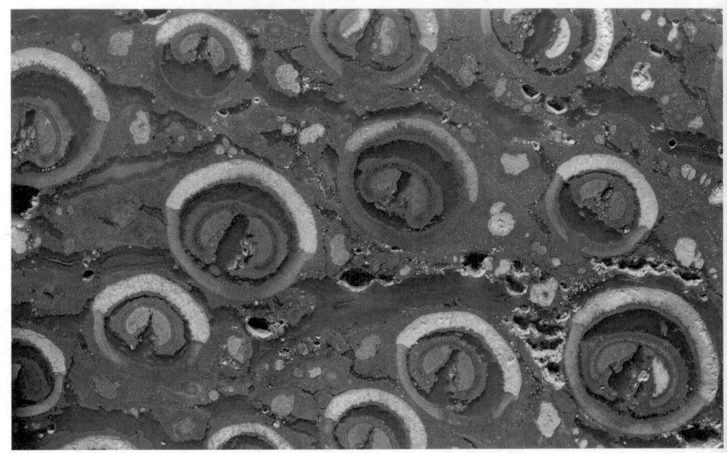

FERN

PETRIFIED WOOD >

LEPIDODENDRON

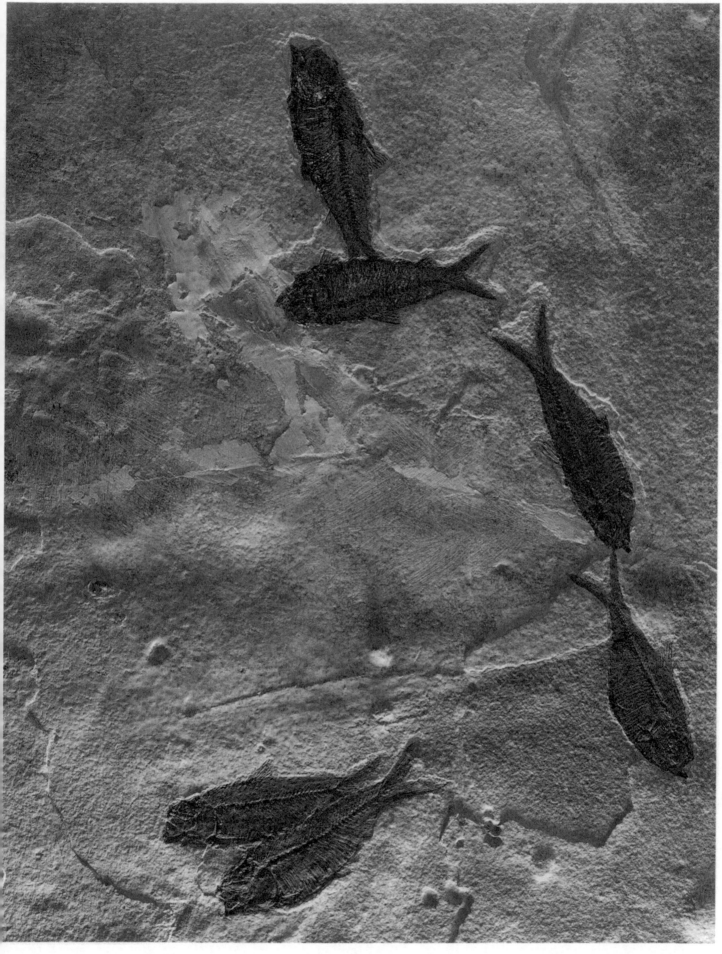

< TRILOBITES

LAST DANCE OF THE FISH

AMMONITE IN THIN SECTION

AMMONITE DECO >

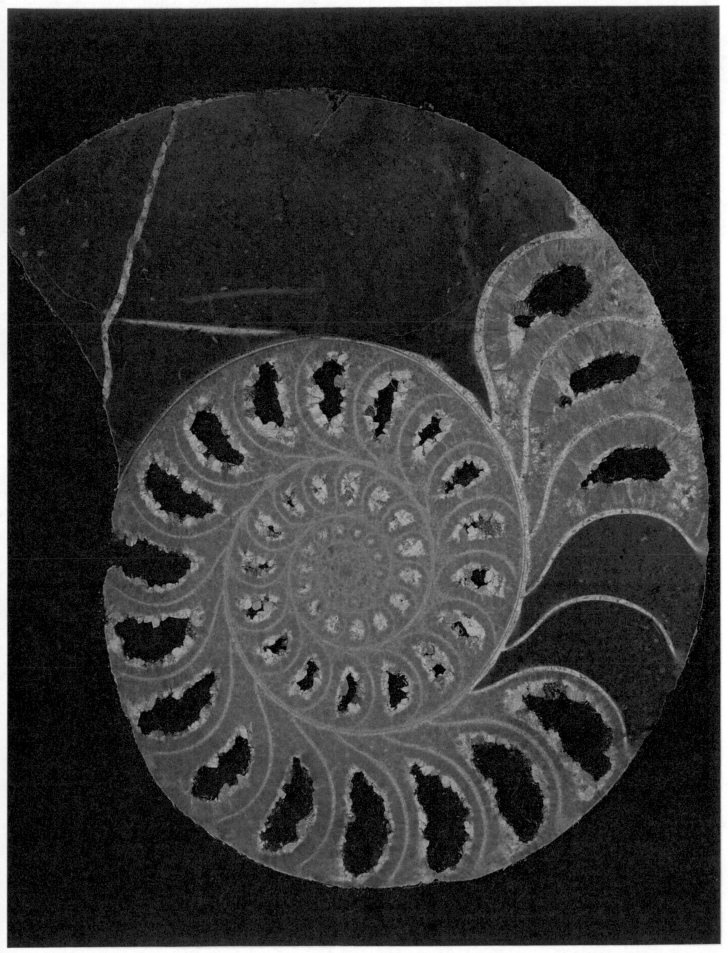

AGATHICERAS TIMORENSIS

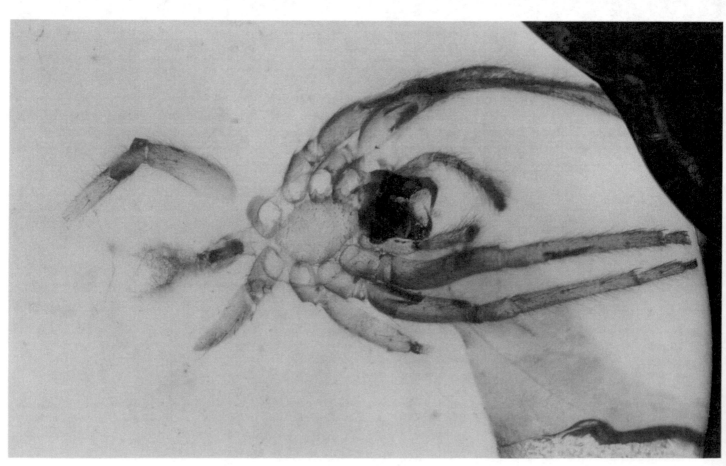

SPIDER LARVA

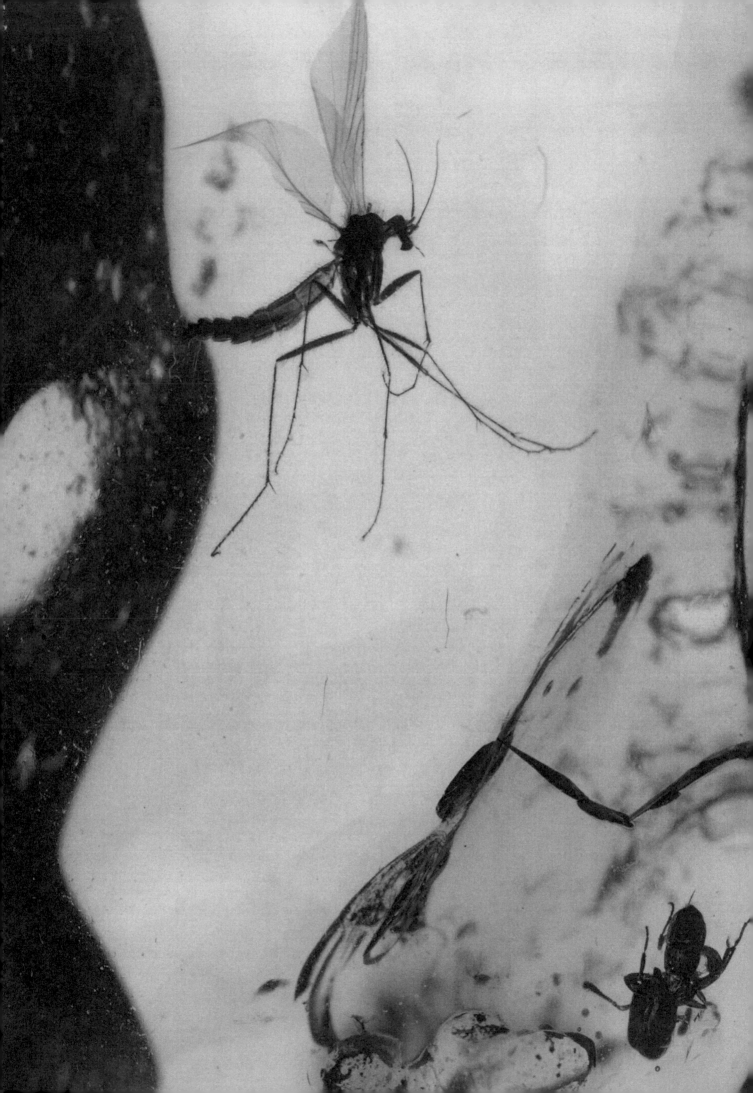

PETRIFIED TREE STUMP

ACKNOWLEDGEMENTS

I owe acknowledgement and thanks to many people, some for their tips and suggestions, others for their readiness to place at my disposal samples from their scientific practice or collection. I would like to thank the following by name:

GEOLOGISTS Charles Arps, Martin Drury, Rob Felius, Armella Kloppenburg, Cees Mayer, Leo Minnigh, Cees Passchier, Harry Priem, Rudolf Trouw, Janos Urai, Jan Werner, Annemarie Wiechowski, Bep van der Wilk and Kees Woensdrecht.

CUTTERS / POLISHERS Karl Kaucher III and his wife Lotte, Inge Nüssgen and Otto Stiekema.

DEALERS AND COLLECTORS of minerals and (semi) precious stones: Richard Abbenhuis, Bert Baasdorp, Dolf Bode, Hans Gregorius, José Hennekam, Roy Masin, Eric Muiderman, Paris Nicolucci, Alfred Peth and Gerard de Roller.

TEXT CORRECTION Madelon Evers.

FOR CRITICAL STIMULATION my daughters Dorine and Caroline, Adriaan Coppens, Susanne Mol and Marianne Wolfs. And last but foremost my wife Johannette, for sharing it all.

LITERATURE CITED Caillois, Roger (1966) 'Pierres'. Gallimard. France. Caillois, Roger (1970) 'l'Écriture des Pierres'. Skira S.A. Genève. Mandelbrot, Benoit B. (1983) 'The Fractal Geometry of Nature'. Freeman, San Francisco. Many other books and publications were consulted.

PHOTOGRAPHY Most of the photographs were taken with a 4 X 5 inch technical camera and a variety of lenses. Most of the field photos were taken with a fixed focus camera (Fuji) on format 60 x 90 mm. All on color reversal film, 50, 64 or 100 ASA, most daylight film and some, notably the thin section recordings, on tungsten film. All photographic recording is by conventional optical means, without digital manipulation or enhancement.

DESIGN Pim Smit, Amsterdam.